Engineering Hydrology

E. M. WILSON
D. Sc., Ph.D., M.Sc., F.I.C.E., F.A.S.C.E.

Research Professor of Hydraulic Engineering
University of Salford

FOURTH EDITION

MACMILLAN

First edition 1969
Second edition 1974
Reprinted eight times
Third edition 1983
Reprinted three times
Fourth edition 1990

Published by
MACMILLAN PRESS LTD
Houndmills, Basingstoke, Hampshire RG21 6XS
and London
Companies and representatives
throughout the world

ISBN 0–333–51716–4 hardcover
ISBN 0–333–51717–2 paperback

A catalogue record for this book is available from the British Library.

14   13   12   11   10   9   8
04   03   02   01   00   99

Printed and bound in Great Britain by
Creative Print and Design (Wales), Ebbw Vale

Engineering Hydrology

Macmillan titles of interest to Civil Engineers

Contents

Preface to the Fourth Edition

This fourth edition, like its three predecessors, is written for engineering students and junior engineers; to introduce them to the principles and practice of engineering hydrology and to show, through many worked examples, how to approach the many apparently intractable problems which hydraulic engineers meet.

The last decade has been a time of considerable activity in the subject, following the publication of the *Flood Studies Report* by the Institute of Hydrology. Examples of this on-going work include: advances in urban hydrology, published as *The Wallingford Procedure*; the *Low Flow Studies*, the *Flood Studies Supplementary Reports* and the *World Flood Study* from IOH; and the *Manual for Estimation of Probable Maximum Precipitation* from the World Meteorological Organisation.

Short descriptions of some of these subjects have been included in this edition to encourage deeper study of the original texts. The opportunity has been taken of enlarging the lists of problems, re-organising several chapters, updating references and including relevant new material.

I continue to receive and much appreciate comments, corrections and advice from correspondents around the world, to whom I am grateful.

Manchester, 1989·   E. M. Wilson

Acknowledgements

Permission to publish copyright material is gratefully acknowledged as follows.

From the Director, the Institute of Hydrology, Wallingford, United Kingdom
  Tables 2.6, 2.7, 2.8, 2.9, 2.11, 4.2 and 6.1
  Figures 2.7, 2.17, 4.9, 4.10, 6.21, 6.24, 7.28, 9.7, 9.8, 11.1, 11.2, 11.3 and 11.4

From the Controller, Her Majesty's Stationery Office
  Appendix A: SAAR, 2DM5, r and RP maps for the United Kingdom
  Figures 2.6, 2.13 and 4.8

From the Soil Survey of England and Wales, the Macauley Institute for Soil Research, the National Soil Survey of Ireland and Mr B. S. Kear
  Appendix A: RP maps for England and Wales, Scotland, Ireland and the Isle of Man respectively

From the Director, Irish Meteorological Service
  Appendix A: SAAR, 2DM5 and r maps of the Republic of Ireland

From the Director, Ghana Meteorological Service
  Figure 2.8

From the Institution of Civil Engineers
  Figures 7.25, 7.26 and 7.27

From Professor L. Huisman, Delft University
  Figure 5.11

From Mr P. J. Rijkoort, Royal Meteorological Institute, The Netherlands
  Appendix C: Nomogram for Penman's equation

From the Cambridge University Press
  Table 3.1

From the American Geographical Society, New York
Figure 3.2

From the McGraw-Hill Book Company
Figure 4.7

From Dr I. G. Littlewood
Figures 6.13 and 6.14

From the American Water Works Association
Figure 2.18

From the Director, Transport and Road Research Laboratory
Tables 10.1 and 10.2
Figure 10.1

From the Director General, U.N. Food and Agriculture Organisation
Tables 3.2, 3.3 and 3.4
Figure 3.4

Grateful acknowledgement of assistance and information is also made to
Mr B. J. Greenfield of Thames Water Authority
Dr Frank Farquharson, Dr I. G. Littlewood and Dr A. Gustard, all of the
Institute of Hydrology
The Meteorological Office Advisory Services
Messrs Boode B. V., Zevenhuizen, The Netherlands for Figure 5.9
Mrs Margaret Pearson, University of Salford

A 64-page booklet containing model answers to all of the numerical questions in this book is available from the publishers.
[ISBN 0–333–52383–0]

1 Introduction

The science of hydrology deals with the occurrence and movement of water on and over the surface of the Earth. It deals with the various forms of moisture that occur, and the transformation between the liquid, solid and gaseous states in the atmosphere and in the surface layers of land masses. It is concerned also with the sea—the source and store of all the water that activates life on this planet.

1.1 Allied sciences

The engineer is normally concerned with the design and operation of engineering works to control the use of water and in particular the regulation of streams and rivers and the formation of storage reservoirs and irrigation canals. Nevertheless, he must be aware of the application of hydrology in the wider context of the subject, since much of his material is derived from physics, meteorology, oceanography, geography, geology, hydraulics and kindred sciences. He must be aware of experience in forestry and agriculture and in botany and biology. He must know probability theory, some statistical methods and be able to use economic analysis.

Hydrology is basically an interpretive science. Experimental work is restricted, by the scale of natural events, to modest researches into particular effects. The fundamental requirement is observed and measured data on all aspects of precipitation, runoff, percolation, river flow, evaporation and so on. With these data and an insight into the many bordering fields of knowledge, the skilled hydrologist can provide the best solutions to many engineering problems that arise.

1.2 The hydrological cycle

The cyclic movement of water from the sea to the atmosphere and thence by precipitation to the Earth, where it collects in streams and runs back to the sea, is referred to as the hydrological cycle. Such a cyclic order of events does

1

occur but it is not so simple as that. Firstly, the cycle may short-circuit at several stages, for example, the precipitation may fall directly into the sea, lakes or river courses. Secondly, there is no uniformity in the time a cycle takes. During droughts it may appear to have stopped altogether, during floods it may seem to be continuous. Thirdly, the intensity and frequency of the cycle depend on geography and climate, since it operates as a result of solar radiation, which varies according to latitude and season of the year. Finally, the various parts of the cycle can be quite complicated and man can exercise some control only on the last part, when the rain has fallen on the land and is making its way back to the sea.

Although this concept of the hydrological cycle is oversimplified, it affords a means of illustrating the most important processes that the hydrologist must understand. The cycle is shown diagrammatically in figure 1.1.

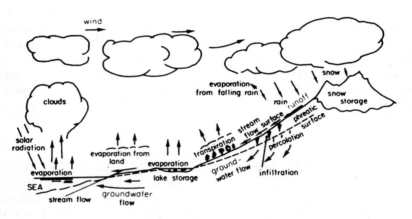

Figure 1.1   *The hydrological cycle*

Water in the sea evaporates under solar radiation, and clouds of water vapour move over land areas. *Precipitation* occurs as snow, hail, rain and condensate in the form of dew, over land and sea. Snow and ice on land are water in temporary storage. Rain falling over land surfaces may be *intercepted* by vegetation and evaporate back to the atmosphere. Some of it *infiltrates* into the soil and moves down or *percolates* into the saturated ground zone beneath the water-table, or phreatic surface. The water in this zone flows slowly through aquifers to river channels or sometimes directly to the sea. The water that infiltrates also feeds the surface plant life and some gets drawn up into this vegetation where *trans-piration* takes place from leafy plant surfaces.

The water remaining on the surface partially evaporates back to vapour, but the bulk of it coalesces into streamlets and runs as surface runoff to the river channels. The river and lake surfaces also evaporate, so still more is removed here. Finally, the remaining water that has not infiltrated or evaporated arrives back at the sea via the river channels. The *groundwater*, moving much more slowly, either emerges into the stream channels or arrives at the coastline and seeps into the sea, and the whole cycle starts again.

1.3 Inventory of Earth's water

It is as well to have a clear idea of the scale of the events that are being discussed. Table 1.1 lists estimates of the amounts of water involved in the hydrological cycle and the proportion (in percentages) of the total water on Earth involved in each part of it.

TABLE 1.1 *Estimated Earth's water inventory*

Location	Volume $(10^3 \ km^3)$	Percentage total water
Fresh-water lakes	125	
Rivers	1·25	
Soil moisture	65	0·62
Groundwater	8250	
Saline lakes and inland seas	105	0·008
Atmosphere	13	0·001
Polar ice-caps, glaciers and snow	29200	2·1
Seas and oceans	1320000	97·25
Total	1360000 or $1·36 \times 10^{18} \ m^3$	100·0

Of the 0·6 per cent of total water that is available as fresh water, about half is below a depth of 800 m and so is not practically available on the surface. This means that the stock of the Earth's fresh water that is obtainable one way or another for man's use is about $4 \times 10^6 \ km^3$ and is mainly in the ground. Spread over the Earth's land surface it would be about 30 m deep.

The four processes with which the hydrologist is mainly concerned are precipitation, evaporation and transpiration, surface runoff or stream flow, and groundwater flow. He needs to be able to interpret data about these processes and to predict from his studies the most likely quantities involved in the extreme cases of flood and drought. He must be able also to express an opinion about the likely frequency with which such events will occur, since it is on the frequency of certain values of extreme events that much hydraulic engineering design is based.

1.4 Hydrology as applied in engineering

To the practising engineer concerned with the planning and building of hydraulic structures, hydrology is an indispensable tool. Suppose, for example, that a city wishes to increase or improve its water supply. The engineer first looks for sources of supply; having perhaps found a clear uninhabited mountain catchment area, he must make an estimate of its capability of supplying water. How much rain will fall on it? How long will dry periods be and what amount of storage will be necessary to even out the flow? How much of the runoff will be lost as evaporation and transpiration? Would a surface storage scheme be better than abstraction of the groundwater flow from wells nearer the city?

The questions do not stop there. If a dam is to be built, what capacity must the spillway have? What diameter should the supply pipelines be? Would afforestation of the catchment area be beneficial to the scheme or not?

To all these questions, and many others that arise, the hydrologist can supply answers. Often the answers will be qualified, and often also they will be given as probable values, with likely deviations in certain lengths of time. This is because hydrology is not an exact science. A contractor may be building a cofferdam in a river and his hydrologist may tell him that, built to a certain height, it will be overtopped only once, on average, in 100 years. If it is a temporary structure built for a service life of maybe only 2 years the contractor may decide this is a fair risk. However, it *is* a risk. One of the 2 years could be that in which the once in 100 years flood arrives, and the science of hydrology cannot, as yet, predict this.

In a broader field of engineering, which is of great and increasing importance, the development of water resources over a whole river basin or geographical region may be under consideration. In these circumstances the role of the hydrologist is especially important. Now his views and experience are of critical weight not only in the engineering structures involved in water supply, but also in the type and extent of the agriculture to be practised, in the siting of industry, in the size of population that can be supported, in the navigation of inland shipping, in port development and in the preservation of amenities.

Civilisation is primarily dependent on water supply. As the trend towards larger cities and increasing industrialisation continues, so will the role of the hydrologist increase in importance in meeting the demands of larger populations for water for drinking, sanitation, irrigation, industry and power generation.

2 Meteorological Data

2.1 Weather and climate

The hydrology of a region depends primarily on its climate, secondly on its topography and its geology. Climate is largely dependent on the geographical position on the earth's surface. Climatic factors of importance are precipitation and its mode of occurrence, humidity, temperature and wind, all of which directly affect evaporation and transpiration.

Topography is important in its effect on precipitation and the occurrence of lakes, marshland and high and low rates of runoff. Geology is important because it influences topography and because the underlying rock of an area is the groundwater zone where the water that has infiltrated moves slowly through aquifers to the rivers and the sea.

The pattern of circulation in the atmosphere is complex. If the earth were a stationary uniform sphere, then there would be a simple circulation of atmosphere on that side of it nearest the sun. Warmed air would rise at the equator and move north and south at a high altitude, while cooler air moved in across the surface to replace it. The high warm air would cool and sink as it moved away from the equator, until it returned to the surface layers when it would move back to the equator. The side of the earth remote from the sun would be uniformly dark and cold.

This simple pattern is upset by the earth's daily rotation, on its own axis, which gives alternate 12-hour heating and cooling and also produces the Coriolis force acting on airstreams moving towards or away from the equator. It is further upset by the tilt of the earth's axis to the plane of its rotation around the sun, which gives rise to seasonal differences. Further effects are due to the different reflectivity and specific heats of land and water surfaces. The result of these circumstances on the weather is to make it generally complex and difficult to predict in the short term. By observations of data over a period of time, however, long-term predictions may be made on a statistical basis.

The study of hydrology necessitates the collection of data on (among other factors) humidity, temperature, precipitation, radiation and wind velocity. All of these main factors are considered in this chapter.

5

2.2 Humidity

Air easily absorbs moisture in the form of water vapour. The amount absorbed depends on the temperatures of the air and of the water. The greater the temperature of the air, the more water vapour it can contain. The water vapour exerts a *partial pressure* usually measured in either *bars* (1 bar = 100 kN/m^2; 1 millibar = 10^2 N/m^2) or mm height of a column of mercury (Hg) (1 mm Hg = 1.33 mbar).

Suppose an evaporating surface of water is in a closed system and enveloped in air. If a source of heat energy is available to the system, evaporation of the water into the air will take place until a state of equilibrium is reached when the air is saturated with vapour and can absorb no more. The molecules of water vapour will then exert a pressure that is known as *saturation vapour pressure*, or e_s, for the particular temperature of the system.

The value of e_s changes with temperature as indicated in table 2.1. These values are also plotted as a curve connecting e_s and temperature ($^\circ$C) in figure 2.1. Referring to figure 2.1, consider what can happen to a mass of atmospheric air P whose temperature is t and whose vapour pressure is e.

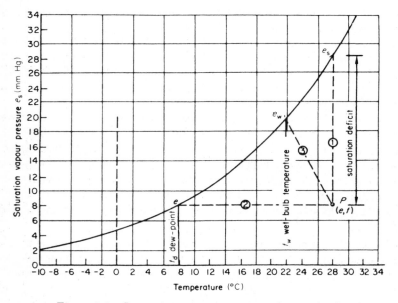

Figure 2.1 *Saturation vapour pressure of water in air*

Since P lies below the saturation vapour pressure curve, it is clear that the air mass could absorb more water vapour and that if it did so while its temperature remained constant, then the position of P would move vertically up dashed line ① until the air was saturated. The corresponding vapour pressure of P in this new position would be e_s. The increase ($e_s - e$) is known as the *saturation deficit*.

TABLE 2.1 *Saturation vapour pressure as a function of temperature t (negative values of t refer to conditions over ice; 1 mm Hg = 1.33 mbar)*

t ($^\circ C$)	e_s (mm Hg)									
	0.0	0.1	0.2	0.3	0.4	0.5	0.6	0.7	0.8	0.9
−10	2.15									
−9	2.32	2.30	2.29	2.27	2.26	2.24	2.22	2.21	2.19	2.17
−8	2.51	2.49	2.47	2.45	2.43	2.41	2.40	2.38	2.36	2.34
−7	2.71	2.69	2.67	2.65	2.63	2.61	2.59	2.57	2.55	2.53
−6	2.93	2.91	2.89	2.86	2.84	2.82	2.80	2.77	2.75	2.73
−5	3.16	3.14	3.11	3.09	3.06	3.04	3.01	2.99	2.97	2.95
−4	3.41	3.39	3.37	3.34	3.32	3.29	3.27	3.24	3.22	3.18
−3	3.67	3.64	3.62	3.59	3.57	3.54	3.52	3.49	3.46	3.44
−2	3.97	3.94	3.91	3.88	3.85	3.82	3.79	3.76	3.73	3.70
−1	4.26	4.23	4.20	4.17	4.14	4.11	4.08	4.05	4.03	4.00
−0	4.58	4.55	4.52	4.49	4.46	4.43	4.40	4.36	4.33	4.29
0	4.58	4.62	4.65	4.69	4.71	4.75	4.78	4.82	4.86	4.89
1	4.92	4.96	5.00	5.03	5.07	5.11	5.14	5.18	5.21	5.25
2	5.29	5.33	5.37	5.40	5.44	5.48	5.53	5.57	5.60	5.64
3	5.68	5.72	5.76	5.80	5.84	5.89	5.93	5.97	6.01	6.06
4	6.10	6.14	6.18	6.23	6.27	6.31	6.36	6.40	6.45	6.49
5	6.54	6.58	6.54	6.68	6.72	6.77	6.82	6.86	6.91	6.96
6	7.01	7.06	7.11	7.16	7.20	7.25	7.31	7.36	7.41	7.46
7	7.51	7.56	7.61	7.67	7.72	7.77	7.82	7.88	7.93	7.98
8	8.04	8.10	8.15	8.21	8.26	8.32	8.37	8.43	8.48	8.54
9	8.61	8.67	8.73	8.78	8.84	8.90	8.96	9.02	9.08	9.14
10	9.20	9.26	9.33	9.39	9.46	9.52	9.58	9.65	9.71	9.77
11	9.84	9.90	9.97	10.03	10.10	10.17	10.24	10.31	10.38	10.45
12	10.52	10.58	10.66	10.72	10.79	10.86	10.93	11.00	11.08	11.15
13	11.23	11.30	11.38	11.75	11.53	11.60	11.68	11.76	11.83	11.91
14	11.98	12.06	12.14	12.22	12.96	12.38	12.46	12.54	12.62	12.70
15	12.78	12.86	12.95	13.03	13.11	13.20	13.28	13.37	13.45	13.54
16	13.63	13.71	13.80	13.90	13.99	14.08	14.17	14.26	14.35	14.44
17	14.53	14.62	14.71	14.80	14.90	14.99	15.09	15.17	15.27	15.38
18	15.46	15.56	15.66	15.76	15.96	15.96	16.06	16.16	16.26	16.36
19	16.46	16.57	16.68	16.79	16.90	17.00	17.10	17.21	17.32	17.43
20	17.53	17.64	17.75	17.86	17.97	18.08	18.20	18.31	18.43	18.54
21	18.65	18.77	18.88	19.00	19.11	19.23	19.35	19.46	19.58	19.70
22	19.82	19.94	20.06	20.19	20.31	20.43	20.58	20.69	20.80	20.93
23	21.05	21.19	21.32	21.45	21.58	21.71	21.84	21.97	22.10	22.23
24	22.27	22.50	22.63	22.76	22.91	23.05	23.19	23.31	23.45	23.60
25	23.75	23.90	24.03	24.20	24.35	24.49	24.64	24.79	24.94	25.08
26	25.31	25.45	25.60	25.74	25.89	26.03	26.18	26.32	26.46	26.60
27	26.74	26.90	27.05	27.21	27.37	27.53	27.69	27.85	28.00	28.16
28	28.32	28.49	28.66	28.83	29.00	29.17	29.34	29.51	29.68	29.85
29	30.03	30.20	30.38	30.56	30.74	30.92	31.10	31.28	31.46	31.64
30	31.82	32.00	32.19	32.38	32.57	32.76	32.95	33.14	33.33	33.52

Alternatively, if no change were to take place in the humidity of the air while it was cooled, then P would move horizontally to the left along line ② until the saturation line was intersected again. At this point P would be saturated, at a new temperature t_d, the *dew-point*. Cooling of the air beyond this point would result in condensation or mist being formed.

If water is allowed to evaporate freely into the air mass, neither of the above two possibilities occurs. This is because the evaporation requires heat, which is withdrawn from the air itself. This heat, called the *latent heat of evaporation*, h_r, is given by the equation

$$h_r = 606.5 - 0.695t \text{ cal/g}$$

So, as the humidity and vapour pressure rise, the temperature of the air falls and the point P moves diagonally along line ③ until saturation vapour pressure is reached at the point defined by e_w and t_w. This temperature t_w is called the *wet-bulb temperature* and is the temperature to which the original air can be cooled by evaporating water into it. This is the temperature found by a wet-bulb thermometer.

The *relative humidity* is now given as

$$h = e/e_s, \text{ or as a percentage}, h = 100 \, e/e_s \text{ per cent}$$

and is a measure of the air's capacity, at its existing temperature, to absorb further moisture. It is measured by blowing air over two thermometers, one with its bulb wrapped in wet muslin and one dry. The air flow past the bulb has an influence on the wet-bulb reading and the two thermometers can either be whirled around on a string or more conveniently have the air current provided by a clockwork fan. In this latter case the instrument is called a *psychrometer*.

The value of e for air temperature t may be obtained from the equation

$$(e_w - e) = \gamma(t - t_w)$$

where t_w = wet-bulb temperature

   t = dry-bulb temperature

   e_w = the corresponding partial pressures for t_w (from table 2.1)

   γ = psychrometer constant (assuming the air speed past the bulbs exceeds 3 m/s and t is measured in °C, then:
   for e in mbar, $\gamma = 0.660$
   for e in mm Hg, $\gamma = 0.485$).

2.3 Temperature

Air temperature is recorded by thermometers housed in open louvred boxes, known as Stevenson screens, about 1.25 m above ground. Protection is necessary from precipitation and the direct rays of the sun.

Many temperature observations are made using *maximum and minimum thermometers*. These record, by indices, the maximum and minimum temperatures experienced since the instrument was last set.

The daily variation in temperature varies from a minimum around sunrise, to a maximum from $\frac{1}{2}$ to 3 hours after the sun has reached its zenith, after which there is a continual fall through the night to sunrise again. Accordingly, maximum and minimum observations are best made in the period from 8 a.m. to 9 a.m. after the minimum has occurred.

The *mean daily temperature* is the average of the maximum and minimum and is normally within a degree of the true average as continuously recorded.

Temperature is measured in degrees Celsius, commonly, though erroneously, called centigrade. The Fahrenheit scale is still also in common use.

Vertical temperature gradient. The rate of change of temperature in the atmosphere with height is called the *lapse rate*. Its mean value is 6.5 °C per 1000 m height increase. This rate is subject to variation, particularly near the surface, which can become very warm by day, giving a higher lapse rate, and cool by night, giving a lower lapse rate. The cooling of the earth, by outward radiation, on clear nights can be such that a *temperature inversion* occurs, with warmer air overlying the surface layer.

As altitude increases, barometric pressure decreases so that a unit mass of air occupies greater volume the higher it rises. The temperature change due to this decompression is about 10 °C per 1000 m if the air is dry. This is the *dry-adiabatic* lapse rate. If the air is moist, then as it is lifted, expanding and cooling, its water vapour content condenses. This releases latent heat of condensation, which prevents the air mass cooling as fast as dry air. The resulting *saturated-adiabatic* lapse rate is therefore lower, at about 5·6 °C per 1000 m in the lower altitudes.

Distribution of temperature. Generally, the nearer to the equator a place is, the warmer that place is. The effects of the different specific heats of earth and water, the patterns of oceanic and atmospheric currents, the seasons of the year, the topography, vegetation and altitude all tend to vary this general rule, and all need consideration.

2.4 Radiation

Most meteorological recording stations are equipped with *radiometers* to measure both incoming short-wave radiation from sun and sky, and net radiation, which is the algebraic sum of all incoming radiation and the reflected short-wave and long-wave radiation from the earth's surface. The net radiation is of great importance in evaporation studies, as will be seen in chapter 3.

2.5 Wind

Wind speed and direction are measured by *anemometer* and wind vane respectively. The conventional anemometer is the *cup anemometer* formed by a circlet of three (sometimes four) cups rotating around a vertical axis. The speed of rotation measures the wind speed and the total revolutions around the axis gives a measure of *wind run*, the distance a particular parcel of air travels in a specified time.

Because of the frictional effects of the ground or water surface over which the wind is blowing, it is important to specify in any observation of wind, the height above ground at which it was taken. An empirical relationship between wind speed and height has been commonly used

$$u/u_0 = (z/z_0)^{0.15}$$

where u_0 = wind speed at anemometer at height z_0

u = wind speed at some higher level z.

In recent years there has been some effort to standardise observation heights and in Europe wind speed is usually observed 2 m above the surface.

Figure 2.2 shows an instrument array for making meteorological observations at regular, short time-intervals. Instruments, which record automatically on magnetic tape, include net-radiation radiometer, wet and dry bulb thermometers, wind vane, anemometer and incident solar radiometer at the mast-top.

2.6 Precipitation

The source of almost all our rainfall is the sea. Evaporation takes place from the oceans and water vapour is absorbed in the air streams moving across the sea's surface. The moisture-laden air keeps the water vapour absorbed until it cools to below dew-point temperature when the vapour is precipitated as rain, or if the temperature is sufficiently low, as hail or snow.

The cause of the fall in temperature of an air mass may be due to convection, the warm moist air rising and cooling to form cloud and subsequently to precipitate rain. This is called *convective precipitation*. This is typified by the late afternoon thunderstorms that develop from day-long heating of moist air, rising into towering anvil-shaped clouds. *Orographic precipitation* results from ocean air streams passing over land and being deflected upward by coastal mountains, thus cooling below saturation temperature and spilling moisture. Most orographic rain is deposited on the windward slopes. The third general classification of rainfall is *cyclonic and frontal precipitation*. When low-pressure areas exist, air tends to move into them from surrounding areas and in so doing displaces low-pressure air upward, to cool and precipitate rain. Frontal rain is associated with the boundaries of air masses where one mass is colder than the other and so intrudes a cool wedge under it, raising the warm air to form clouds and rain. The slope of these frontal wedges can be quite flat and so rain areas associated with fronts may be very large.

2.6.1 Recording precipitation.
Precipitation occurs mainly as rain, but can occur also as hail, sleet, snow, fog or dew. Britain has a humid climate and rain provides the great bulk of its moisture, but in other parts of the world precipitation can be almost entirely snow, or, in arid zones, dew.

In the United Kingdom, rainfall records are received and recorded by the

Figure 2.2 *Meteorological observation array. On lower arm left, net radiation; right, wet and dry bulb thermometers. On upper arm left, wind direction; right, wind run. At top, solar and sky radiation. There is a rain gauge with anti-splash screen in the middle distance*

Meteorological Office from some 6500 rain gauges scattered over Great Britain and Northern Ireland, the majority giving daily values of rainfall. In addition there are a further 260 stations equipped also with recording rain gauges that record continuously.

Standard rain gauges in Britain are made from copper and consist of a 5-in. diameter copper cylinder, with a chamfered upper edge, which collects the rain and allows it to drain through a funnel into a removable container of metal or glass from which the rain may be poured into a graduated glass measuring-cylinder each day. There are prescribed patterns for the standard gauge and for its installation and operation.

Recording gauges (or autographic rain recorders) usually work by having a clockwork-driven drum carrying a graph on which a pen records either the total weight of container plus water collected, or a series of blips made each time a small container of known capacity spills its contents. Such gauges are more expensive and more liable to error but may be the only kind possible for remote, rarely visited sites. They have the great advantage that they indicate *intensity* of rainfall, which is a factor of importance in many problems. For this reason some stations are equipped with both standard and recording gauges.

The Meteorological Office has recently designed a new range of rain gauges [1]. The new standard gauge for daily rainfall measure is a circular catchment of 150 cm^2 (5·5 in. diameter) installed at a rim height of 300 mm above ground level. The larger and more accurate new gauge has an area of 750 cm^2 (12·2 in. diameter) also set with a rim 300 mm above ground. The material used in manu-facture is fibre glass. A new tipping-bucket mechanism has been designed and is available with a telemetry system to provide for distant reading by telephone interrogation. The gauge is provided with a telephone connection and number, which can be dialled in the ordinary way. The quantity of rain collected since the last setting of the gauge to zero is transmitted in increments of 1 mm by three groups of audible tones representing hundreds, tens and units. Interroga-tion can be made as frequently as desired and hence intensities can be obtained, by simple subtraction, with minimum delay [2].

In recent years there has been much research into the effects of exposure on rain gauges and it is now generally accepted that more accurate results will be obtained from a rain gauge set with its rim at ground level, than one with its rim some height above ground [3]. It is necessary in a ground-level installation to make a pit to house the gauge and cover it with an anti-splash grid. Accordingly the ground-level gauge is more expensive to install and maintain.

A typical installation is illustrated in figure 2.3.

Yearly records for the whole country, statistically analysed and presented graphically, are published annually by the Meteorological Office in a booklet entitled *British Rainfall* followed by the particular year concerned. The use of rainfall data is discussed in section 2.8.

2.6.2 Rain-gauge networks. A question that frequently arises concerns the number and type of rain gauges that are necessary to ensure an accurate assess-

Figure 2.3 *Automatic recording rain gauge set with rim at ground level and with anti-splash screen*

ment of a catchment's rainfall. Bleasdale [4] quotes tables 2.2 and 2.3 as general guides and comments as follows

The disparity between the two tables is not so great as might appear at first sight. The first indicates station densities which are reached in important reservoired areas and which may well be exceeded in small experimental areas. The

TABLE 2.2   *Minimum numbers of rain gauges required in reservoired moorland areas*[a]

Area		Rain gauges		
mile²	*km²*	*Daily*	*Monthly*	*Total*
0.8	2	1	2	3
1.6	4	2	4	6
7.8	20	3	7	10
15.6	41	4	11	15
31.3	81	5	15	20
46.9	122	6	19	25
62.5	162	8	22	30

[a]Source: Joint Committee of the Meteorological Office, Royal Meteorological Society, and the Institution of Water Engineers Report on 'The determination of the general rainfall over any area'. *Trans. Institution of Water Engineers*, **42** (1937) 231.

TABLE 2.3   *Minimum numbers of rain gauges for monthly percentage-of-average rainfall estimates* [4]

Area		Number of rain gauges
mile²	*km²*	
10	26	2
100	260	6
500	1300	12
1000	2600	15
2000	5200	20
3000	7800	24

second indicates densities which are more appropriate for country-wide networks. In the application of the general guidance embodied in this Table [2.3] it must be appreciated that any large river basin will almost invariably have within it a number of sub-basins for which the relatively dense networks would be recommended. Moreover, the minimum densities suggested would often be substantially increased in mountainous areas, and would be closely followed only in areas of low or moderate elevation without complex topography.

There is considerable material devoted to this question of hydrological network design and the reader is referred to the further reading at the end of the chapter.

2.7 Forms of precipitation other than rain

Snow and ice. Snow has the capacity to retain water and so acts as a form of storage. Its density and, therefore the quantity of water contained, varies from as little as 0.005 for newly fallen snow to as much as 0.6 in old, highly compressed snow. Since density varies with depth, samples must be taken at various horizons in a snow pack before the water equivalent can be computed. This is usually done with a sampling tube.

Snowfall may be measured directly by an ordinary rain gauge fitted with a heating system, or by a simple snow stake if there is no drifting, and density is determined simultaneously.

Snow *traverses* are made as field surveys along lines across catchments, to determine snow thicknesses and densities at depth so that water equivalents can be calculated for flood forecasts.

Fog. Estimates of amounts of moisture reaching the ground from fog formation have been made by installing fog collectors over standard rain gauges. Collectors consist of wire gauge cylinders on which moisture droplets form and run down into the rain gauge. Comparisons with standard rain-gauge records at the same locality show differences that are a measure of fog precipitation. The interpretation of such data requires experience and the use of conversion factors, but can make substantial differences (of the order of 50 to 100 per cent) to precipitation in forest areas.

Dew. Dew collectors have been used in Sweden and Israel to measure dew fall. They were made as conical steel funnels, plastic coated and with a projected plan area of about 1 square metre. Dew ponds are used as a source of water in some countries. They are simply shallow depressions in the earth lined with ceramic tiles.

Condensation. Although fog and dew are condensation effects, condensation also produces precipitation from humid air flows over ice sheets and in temperate climates by condensation in the upper layers of soil. Such precipitation does not occur in large amounts but may be sufficient to sustain plant life.

2.8 The extension and interpretation of data

2.8.1 Definitions. The total annual amount of rain falling at a point is the usual basic precipitation figure available. For many purposes, however, this is not adequate and information may be required under any or all of the following headings.

    (i) *Intensity.* This is a measure of the quantity of rain falling in a given time; for example, mm per hour.

    (ii) *Duration.* This is the period of time during which rain falls.

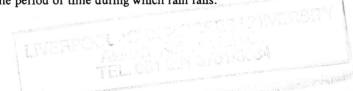

(iii) *Frequency.* This refers to the expectation that a given depth of rainfall will fall in a given time. Such an amount may be equalled or exceeded in a given number of days or years.

(iv) *Areal extent.* This concerns the area over which a point's rainfall can be held to apply.

2.8.2 Intensity—duration relationship. The greater the intensity of rainfall, in general the shorter length of time it continues. A formula expressing the connection would be of the type

$$i = \frac{a}{t + b}$$

where i = intensity (mm/h) t = time (h) a and b are locality constants, and for durations greater than two hours

$$i = \frac{c}{t^n}$$

where c and n are locality constants.

The world's highest recorded intensities are of the order of 40 mm (or $1\frac{1}{2}$ in.) in a minute, 200 mm (or 8 in.) in 20 minutes and 26 m (or 1000 in.) in a year. More detailed information is provided in table 2.4 below.

TABLE 2.4 *World's greatest recorded point rainfalls*

Duration		Depth		Location	Year
		in.	mm		
	1 min	1.5	38	Guadeloupe	1970
	8 min	5.0	126	Bavaria	1920
	20 min	8.1	206	Romania	1889
	42 min	12.0	305	Missouri, USA	1947
2 h	45 min	22.0	559	Texas, USA	1935
12 h		52.8	1340	Réunion (Indian Ocean Island)	1964
24 h		73.6	1870	Réunion	1952
2 days		98.4	2500	Réunion	1952
4 days		146.5	3721	Cherrapunji, India	1974
8 days		162.6	4130	Réunion	1952
1 month		366.1	9300	Cherrapunji	1861
6 months		884.0	22454	Cherrapunji	1861
1 year		1041.8	26461	Cherrapunji	1861

Paulhus [5] suggests that if rainfall is plotted against duration, both scales being logarithmic, the world's greatest recorded rainfalls lie on or just under a straight line whose equation is

$$R = 16.6 \, D^{0.475}$$

where R is rainfall (in.) and D is duration (h), or

$$R = 422D^{0.475}$$

where R is in mm.

Apparently British maxima also lie close to a straight line on a similar plot with values close to one-quarter of the world's values.

British data are presented in table 2.5 and are plotted in figure 2.4. Maximum recorded rainfall in the British Isles is closely approximated by

$$R = 106 \, D^{0.46}$$

where R is rainfall (mm) and D is time period (h).

TABLE 2.5 *Extreme rainfall events in the British Isles*[a]

Duration		Depth (mm)	Location	Year[b]
	1 min	5.1	Croydon, London	1935
	4 min	12.7	Ilkley, West Yorkshire	1906 (a)
	12 min	50.8	Wisbech, Cambridgeshire	1970 (b)
	20 min	63.5	Sidcup, Kent	1958
	45 min	97.0	Orra Beg, Northern Ireland	1980 (c)
1 h		110.2	Wheatley, Oxfordshire	1910
1 h	45 min	154.7[c]	Hewenden, Yorkshire	1956 (d)
2 h		140	Hampstead, London	1975
3 h		178	Horncastle, Lincolnshire	1960 (e)
8 h		200	Bruton, Somerset	1917
24 h		279	Martinstown, Dorset	1955
2 days		300	Sloy, Strathclyde	1974 (f)
4 days		329	Sloy, Strathclyde	1974
12 days		556	Honister Pass, Cumbria	1978
1 month		1436	Llyn Llydaw, Gwynedd	1909 (g)
1 year		6528	Sprinkling Tarn, Cumbria	1954 (h)

[a]Source: S. D. Burt, Meteorological Office, Bracknell, United Kingdom. Maps of average annual rainfall for the period 1941–1970 are reproduced (entitled SAAR) in appendix A, for all of the British Isles except some of the more remote Scottish islands.

[b]Bracketed letters refer to figure 2.4.

[c]Some authorities are doubtful about this figure.

2.8.3 Intensity–duration–frequency relationships. In 1935 Bilham published his well-known article on these relationships in the United Kingdom [6], which contained a graph that is reproduced here as figure 2.5. This graph used the subjective phraseology of 'very rare', 'remarkable' and 'noteworthy' rather than frequency of occurrence. However, the frequencies were calculable from the formula

$$n = 1.25t \, (r + 0.1)^{-3.55}$$

where n = number of occurrences in 10 years

   r = depth of rain in inches

   t = duration of rain in hours.

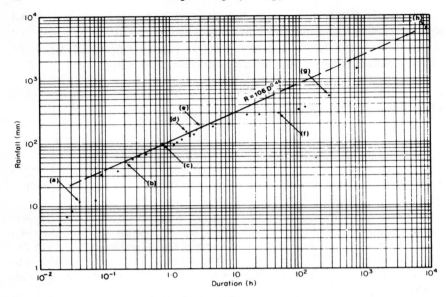

Figure 2.4   *Greatest recorded point rainfalls in the British Isles*

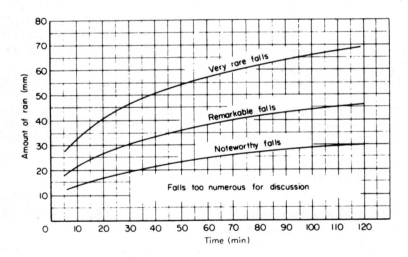

Figure 2.5   *Bilham's rainfall classification*

The SI form of the formula is

$$n = 0.5t \, (P + 2.54)^{-3.55} \tag{2.1}$$

where P = depth of rain in mm, and n and t have unchanged units.

Bilham's work was revised and extended by Holland [8] who showed that Bilham's equations overestimate the probabilities of high-intensity rainfall; that is, above about 35 mm/h. This later work is best illustrated graphically and figure 2.6 shows both Bilham's formulae (chain-dotted lines) and Holland's revisions (full lines). The figure gives a return period for specific depths of rain occurring in specific periods of time, as averaged over 14 station-decades in England.

Another way of presenting such data, this time based on the correlation between annual average rainfall in Britain and one-day maximum rain-depth for various return periods is shown in figure 2.7.

For a specific locality it is often possible to produce curves such as those shown in figure 2.8 for Oxford, England, and Kumasi, Ghana. The difference between coastal temperate and tropical climates should be noted.

The data may also be presented in the form of maps of a region, with iso-hyetal lines indicating total rainfall depth that may be expected in a time t, at a frequency of once in N years. A classic publication of this type by Yarnall [9] shows such maps for the USA. Figure 2.9 is a typical one, reproduced from Yarnall's paper and showing the five-minute rainfall that may be expected once in 50 years.

2.8.4 Depth–area–time relationships.

Precipitation rarely occurs uniformly over an area. Variations in intensity and total depth of fall occur from the centres to the peripheries of storms [10]. The form of variation is illustrated in figure 2.10, which shows for a particular storm, how the average depth of fall decreases from the maximum as the considered area increases.

It is useful, however, to quantify this and Holland [11] has shown that the ratio between point and areal rainfall over areas up to 10 km^2 and for storms lasting from 2 to 120 minutes is given by

$$\frac{\bar{P}}{P} = 1 - \frac{0 \cdot 3\sqrt{A}}{t^*} \tag{2.2}$$

where $\bar{P}$ = average rain depth over the area

P = point rain depth measured at the centre of the area

A = the area in km^2

t^* = an 'inverse gamma' function of storm time obtained from the correlation in figure 2.11.

Example 2.1. What is the average rainfall intensity in Britain over an area of 5 km^2 during a 60-minute storm with a frequency of once in 10 years?

From figure 2.6, the frequency line of once in 10 years just cuts the 25 mm depth at about 1 hour; hence $P = 25$ mm. Given $t = 60$ minutes, then from figure 2.11, $t^* = 5 \cdot 6$ so

$$\frac{\bar{P}}{P} = 1 - \frac{0 \cdot 3\sqrt{5}}{5 \cdot 6} = 1 - 0 \cdot 12 = 0 \cdot 88$$

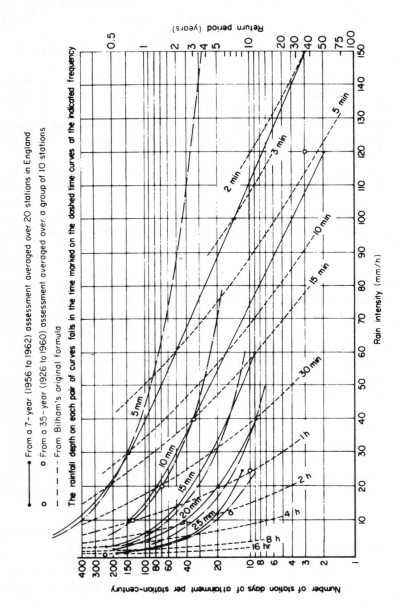

Figure 2.6 *Rain-intensity-frequency graph* [8]. *Example, a rainfall of 20 mm falling in a 30-minute period may be expected, anywhere in Britain, on average once in 11.1 years or 9 times a century*

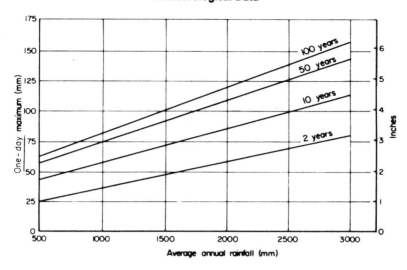

Figure 2.7 *Relationship between one-day maximum rainfall for given return period and average annual rainfall in the United Kingdom (after Institute of Hydrology)*

Hence

$$\bar{P} = 0.88P = 0.88 \times 25 = 22 \text{ mm in 1 hour}$$

Check from Oxford's figures: from figure 2.8 the 10-year frequency for 60-minutes duration indicates 21 mm in 1 hour. Therefore

$$\bar{P} = 0.88P = 0.88 \times 21 = 18.5 \text{ mm in 1 hour}$$

This reduction in average intensity with increase in area, or *areal reduction factor* was considered in Volume II of the Flood Studies Report [13] and subsequently [14], using more extensive data than were available to Holland. For consistency, a further note about it is placed in section 2.9, in a general discussion of the FSR meteorology.

2.8.5 Averaging precipitation depth over an area. In the assessment of total quantities of rainfall over large areas, the incidence of particular storms and their contribution to particular gauges is unknown, and it is necessary to convert many point values to give an average rainfall depth over a certain area. The simplest way of doing this is to take the arithmetical mean of the amounts known for all points in the area. If the distribution of such points over the area is uniform and the variation in the individual gauge's amounts are not large, then this method gives reasonably good results.

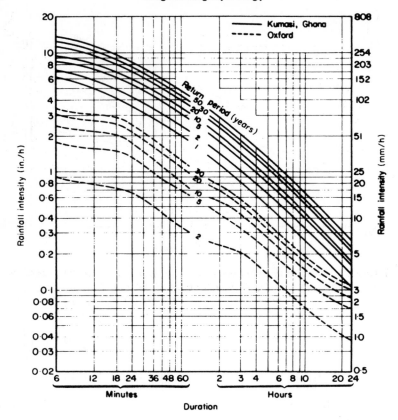

Figure 2.8　*Rainfall frequency–intensity–duration relationships taken at Kumasi, Ghana (courtesy of Ghanaian Meteorological Service) and Oxford (courtesy of Institute of Hydrology)*

Another method, due to Thiessen [15], defines the zone of influence of each station by drawing lines between pairs of gauges, bisecting the lines with perpendiculars, and assuming all the area enclosed within the boundary formed by these intersecting perpendiculars has had rainfall of the same amount as the enclosed gauge.

A variation of this technique is to draw the perpendiculars to the lines joining the gauges at points of median altitude, instead of at mid length. This *altitude-corrected* analysis is sometimes held to be a more logical approach but as a rule produces little difference in result. Either method is more accurate than that of the simple arithmetic mean but involves much labour. Thiessen polygons are illustrated in figure 2.12.

A third method is to draw *isohyets*, or contours of equal rainfall depth. The areas between successive isohyets are measured and assigned an average value of

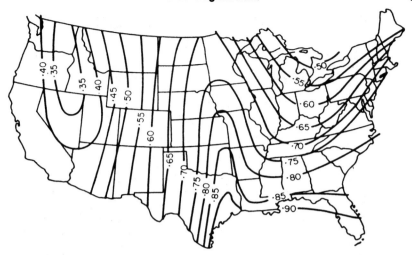

Figure 2.9 *Five minute rainfall, in inches, to be expected once in 50 years in continental USA* [9]

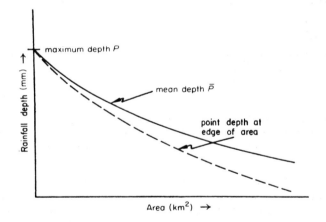

Figure 2.10 *Depth-area curves of rainfall*

rainfall. The overall average for the area is thus derived from weighted averages. This method is possibly the best of the three and has the advantage that the isohyets may be drawn to take account of local effects like prevailing wind and uneven topography. A typical isohyetal map is shown in figure 2.13, though the fall recorded is far from typical, this being the heaviest recorded daily fall in the United Kingdom.

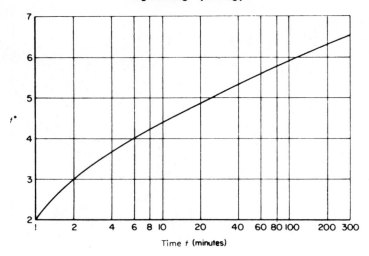

Figure 2.11   *Correlation between storm t and t** [12]

2.8.6 Supplementing rainfall records. It frequently happens when assembling rainfall data that there are areas inadequately recorded, particularly regarding intensities of rainfall. For example, suppose that at two rainfall stations A and B, there is a recording gauge at A and a non-recording gauge at B. Suppose the mass curve of rainfall at A is as shown in the full line on the graph of figure 2.14. The total rainfall at B is known and appears as a point on this graph. If the physical location of B is near A and its rainfall is likely to be of the same kind and frequency, then it is permissible to assume that the mass curve of B will be as shown by the dashed line on the graph. This kind of extension of data should be used with care but can be very useful.

Another example of this is the filling in of a gap in a station's records, when those for neighbouring stations have provided data for the missing period. Suppose for a certain year there is no record of the precipitation at A. In the same year the total at B has been 650 mm.

Assuming that the mean annual precipitation at A and B is 700 mm and 600 mm respectively, then by simple proportion, assuming the average relationship holds for the missing year also, precipitation at A for the missing years will be 700/600 × 650 = 758 mm. This result can be checked by a similar reference to station C.

2.8.7 Apparent trends in observed data. From several years' records it may seem that annual rainfall is, say, declining. It is important to know that this trend is independent of the gauging, and is due to meteorological conditions

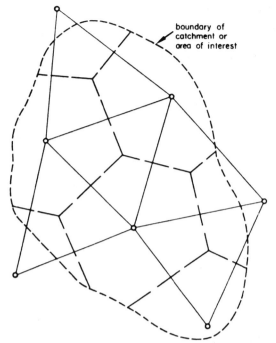

boundary of
catchment or
area of interest

The area assumed to have the rainfall of a particular
gauging station is enclosed by the dashed lines and
catchment boundary

Figure 2.12   *Thiessen polygons*

only. This may be checked by plotting a double mass curve as shown in figure 2.15.

A sudden divergence from the straight-line correlation, shown by the dashed line in the figures, indicates that a change has occurred in gauging and that the meteorology of the region is probably not the cause of the decline. Such a change might be due to the erection of a building or fence near the gauge, which changes the wind pattern round the gauge, the planting of trees, the replacement of one measuring vessel by another, even the changing of an observer to one who uses different procedures.

2.8.8 Trends from progressive averages. Trends can be more clearly discerned by the use of the simple statistical technique of examining averages over longer periods, and moving the group averaged one-year at a time.

Suppose the rainfall records at a station over a number of years are as shown in figure 2.16. The first five years on the record are averaged and this average

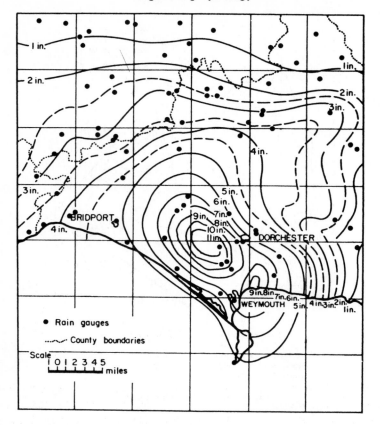

Figure 2.13   *Rainfall over part of Dorset, 18 July 1955 (reproduced from British Rainfall 1955. HMSO, London, 1957)*

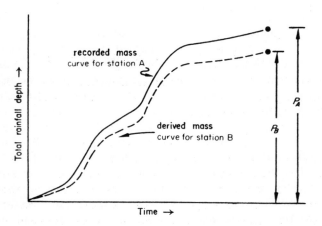

Figure 2.14   *Derivation of rainfall data*

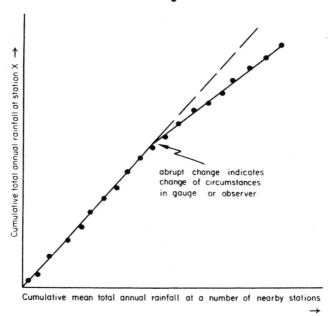

Figure 2.15 *Station check by double mass curve*

is plotted at the mid point of the group. The next point is obtained by omitting the first year and averaging years 2 to 6 inclusive, again plotting the average at the mid point of the group. In this way the wide variations of particular years are smoothed out and long-term trends may be detected.

The same techniques can be applied to temperature, hours of sunshine, wind speeds, cloud cover and other data.

2.9 The meteorological section of the Flood Studies Report (1975)

After the general description of rainfall measurement and interpretation in the preceding sections, there follows a necessarily brief summary of the work undertaken by the United Kingdom Meteorological Office and published in the Flood Studies Report (FSR) [13]. The parameters, derived in the ways described below, are used again in chapters 7 and 9 in sections about flood estimation.

The FSR (Volume II: Meteorology) presents analyses of extensive arrays of British data in maps, tables and graphs. A method is described whereby estimates of rainfall depth at any point in the British Isles, of a given duration and frequency of occurrence (or *return period*), may be made. Means are provided for changing such point rainfalls into areal ones and for selecting appropriate storm profiles (or time distribution of the rain).

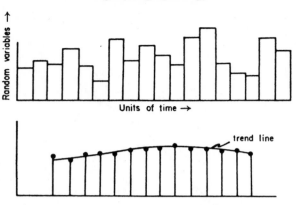

Figure 2.16 *Trends from progressive averages*

2.9.1 Frequency. M5 is adopted as the standard (or reference) frequency. M5 is the depth of rainfall with a return period of 5 years (that is, on average it will be equalled or exceeded once in 5 years). M5 can have a series of different durations: for example, there are 1-minute M5 and 25-day M5 values.

Once the M5 value for a particular point and a particular duration is identified, then the value of MT, where T may have any numerical value of time in years, from 0·5 to 10 000, can also be estimated.

The ratio MT/M5 is called the *growth factor*. Values of growth factor vary slightly geographically and are given in the study for two regions: (a) England and Wales, and (b) Scotland and Northern Ireland. Tables 2.6 and 2.7 provide these values. Growth factors are apparently independent of rainfall duration.

2.9.2 Duration. Here again standards are selected. The standard durations are 2-days and 60-minutes. 2-day M5 rainfall is mapped in the FSR for the British Isles using values from 6000 stations: 60-minute M5 rainfall is mapped for the British Isles as a ratio r = 60-minute M5/2-day M5. Maps of both 2-day M5 and r for almost all of Britain and Ireland are reproduced in appendix A.

2.9.3 Areal reduction factor (ARF). This is the factor that, when applied to point rainfall of specified duration and return period, gives the areal rainfall for the same duration and return period. ARF does not vary appreciably with return period and appears to vary only with area and duration. It was found to be the same for a wide range of geographical locations through variations in duration from 1 minute to 25 days, and in areas from 1 to 30 000 km^2. ARF is tabulated in table 2.8.

TABLE 2.6 *Growth factors MT/M5 for England and Wales* [13]

M5 (mm)	Partial duration series		Annual maximum series						
	2M	1M	M2	M10	M20	M50	M100	M1000	M10 000
0.5	0.52	0.67	0.76	1.14	1.30	1.51	1.70	2.52	3.75
2	0.49	0.65	0.74	1.16	1.32	1.53	1.74	2.60	3.94
5	0.45	0.62	0.72	1.18	1.35	1.56	1.79	2.75	4.28
10	0.43	0.61	0.70	1.21	1.41	1.65	1.91	3.09	5.01
15	0.46	0.62	0.70	1.23	1.44	1.70	1.99	3.32	5.54
20	0.50	0.64	0.72	1.23	1.45	1.73	2.03	3.43	5.80
25	0.52	0.66	0.73	1.22	1.43	1.72	2.01	3.37	5.67
30	0.54	0.68	0.75	1.21	1.41	1.70	1.97	3.27	5.41
40	0.56	0.70	0.77	1.18	1.37	1.64	1.89	3.03	4.86
50	0.58	0.72	0.79	1.16	1.33	1.58	1.81	2.81	4.36
75	0.63	0.76	0.81	1.13	1.27	1.47	1.64	2.37	3.43
100	0.64	0.78	0.83	1.12	1.24	1.40	1.54	2.12	2.92
150	0.64	0.78	0.84	1.11	1.21	1.33	1.45	1.90	2.50
200	0.64	0.78	0.84	1.10	1.20	1.30	1.40	1.79	2.30
500	0.65	0.79	0.85	1.09	1.15	1.20	1.27	1.52	—
1000	0.66	0.80	0.86	1.07	1.12	1.18	1.23	1.42	—

TABLE 2.7 *Growth factors MT/M5 for Scotland and Northern Ireland* [13]

M5 (mm)	Partial duration series		Annual maximum series						
	2M	1M	M2	M10	M20	M50	M100	M1000	M10 000
0.5	0.55	0.68	0.76	1.14	1.30	1.51	1.71	2.54	3.78
2	0.55	0.68	0.76	1.15	1.31	1.54	1.75	2.65	4.01
5	0.54	0.67	0.76	1.16	1.34	1.62	1.86	2.94	4.66
10	0.55	0.68	0.75	1.18	1.38	1.69	1.97	3.25	5.36
15	0.55	0.69	0.75	1.18	1.38	1.70	1.98	3.28	5.44
20	0.56	0.70	0.76	1.18	1.37	1.66	1.93	3.14	5.12
25	0.57	0.71	0.77	1.17	1.36	1.64	1.89	3.03	4.85
30	0.58	0.72	0.78	1.17	1.35	1.61	1.85	2.92	4.60
40	0.59	0.74	0.79	1.16	1.33	1.56	1.77	2.72	4.16
50	0.60	0.75	0.80	1.15	1.30	1.52	1.72	2.57	3.85
75	0.62	0.77	0.82	1.13	1.26	1.45	1.62	2.31	3.30
100	0.63	0.78	0.83	1.12	1.24	1.40	1.54	2.12	2.92
150	0.64	0.79	0.84	1.10	1.20	1.33	1.45	1.90	2.50
200	0.65	0.80	0.85	1.09	1.18	1.30	1.40	1.79	2.30
500	0.66	0.80	0.86	1.08	1.14	1.20	1.27	1.52	—
1000	0.66	0.80	0.86	1.07	1.12	1.18	1.23	1.42	—

TABLE 2.8　*Areal reduction factor (ARF)* [13]

Duration D	Area A (km^2)									
	1	5	10	30	100	300	1000	3000	10 000	30 000
1 min	0.76	0.61	0.52	0.40	0.27	–	–	–	–	–
2 min	0.84	0.72	0.65	0.53	0.39	–	–	–	–	–
5 min	0.90	0.82	0.76	0.65	0.51	0.38	–	–	–	–
10 min	0.93	0.87	0.83	0.73	0.59	0.47	0.32	–	–	–
15 min	0.94	0.89	0.85	0.77	0.64	0.53	0.39	0.29	–	–
30 min	0.95	0.91	0.89	0.82	0.72	0.62	0.51	0.41	0.31	–
60 min	0.96	0.93	0.91	0.86	0.79	0.71	0.62	0.53	0.44	0.35
2 h	0.97	0.95	0.93	0.90	0.84	0.79	0.73	0.65	0.55	0.47
3 h	0.97	0.96	0.94	0.91	0.87	0.83	0.78	0.71	0.62	0.54
6 h	0.98	0.97	0.96	0.93	0.90	0.87	0.83	0.79	0.73	0.67
24 h	0.99	0.98	0.97	0.96	0.94	0.92	0.89	0.86	0.83	0.80
48 h	–	0.99	0.98	0.97	0.96	0.94	0.91	0.88	0.86	0.82
96 h	–	–	0.99	0.98	0.97	0.96	0.93	0.91	0.88	0.85
192 h	–	–	–	0.99	0.98	0.97	0.95	0.92	0.90	0.87
25 days	–	–	–	–	0.99	0.98	0.97	0.95	0.93	0.91

2.9.4 The use of the method.　To obtain point and areal rainfall for any chosen location, duration and return period, the procedure is as follows.

(a) Identify the point by its National Grid Reference (NGR). The National Grid is based on a network of 100 km squares. A diagram showing these and the letters used to designate them is shown in appendix A. A similar diagram for the Irish Grid is provided for the Irish section of appendix A. Most maps used by hydrologists will be 1:25 000 or 1:50 000 series, which use these letters. FSR maps use only the 100 km square numbers. To give an NGR, read the western north–south line number of the square where the point lies, and estimate or measure tenths and hundredths east of it: followed by the southern east–west line, and estimate or measure tenths and hundredths north of it.

(b) Determine the corresponding location on the maps of 2-day M5 and *r* and extract values.

(c) Using table 2.9 (which is a model for M5 rainfall for various durations), and interpolating as necessary for *r*, the values of M5 at the chosen duration, expressed as a percentage of 2-day M5, can be found.

(d) With this value of M5 (mm), use the appropriate growth factor (table 2.6 or 2.7) to establish the chosen duration M*T* factor. The application of this factor to the M5 value gives the point rainfall of chosen duration and *T*-years return period.

(e) Now use the ARF (table 2.8) to establish the areal rainfall for the appropriate area around the chosen point.

TABLE 2.9 *Model for M5 rainfall for durations up to 48 hours* [13]

r (per cent)	M5 rainfall (amounts as percentages of 2-day M5)												
	1 min	2 min	5 min	10 min	15 min	30 min	60 min	2 h	4 h	6 h	12 h	24 h	48 h
12	0.8	1.4	2.7	4.2	5.4	8.1	12	18	26	33	49	72	106
15	1.2	2.1	3.8	5.8	7.2	10.5	15	21	30	37	53	75	106
18	1.6	2.8	5.0	7.4	9.2	12.9	18	25	34	41	56	77	106
21	2.1	3.5	6.3	9.2	11.2	15.5	21	28	38	45	60	80	106
24	2.5	4.3	7.6	11.0	13.3	18.1	24	31	41	48	63	82	106
27	3.0	5.0	9.0	12.9	15.5	20.7	27	35	44	51	65	83	106
30	3.3	5.7	10.3	14.8	17.7	23.3	30	38	48	55	68	85	106
33	3.8	6.5	11.7	16.7	19.9	26.0	33	41	51	57	71	87	106
36	4.1	7.2	13.0	18.6	22.2	28.7	36	44	54	60	73	88	106
39	4.6	8.0	14.5	20.6	24.5	31.5	39	47	57	63	75	89	106
42	5.0	8.7	16.0	22.7	26.9	34.2	42	50	60	66	77	91	106
45	5.4	9.5	17.4	24.7	29.2	37.0	45	53	63	68	79	92	106

An example will illustrate the use of the method more clearly.

Example 2.2. Estimate the 100-year return period rainfall of 6-h duration over a catchment area of 75 km^2 surrounding the NGR point NM950700.

This is a mountainous region of Argyllshire in Western Scotland. NM is the 100 km square between 700 and 800N and 100 and 200E of origin. This lies in section 7 of the key map of Britain.

(i) From appendix A, 2DM5.7 and *r*.7, establish 2-day M5 and *r*.
    2-day M5 = 145 mm; *r* = 14.
(ii) Now use table 2.9 to find M5 for 6-h at *r* = 14.
    6-h M5 (mm) = 36 per cent of 2DM5 = 52 mm.
(iii) Table 2.7 now provides the growth factor to determine M100 (that is, M100/M5 for 52 mm). Growth factor = 1.71 so 6-h M*T* (where *T* = 100 years) = 1.71 (52) = 89 mm.
(iv) Now use table 2.8 for 6 h and 75 km^2. Since ARF = 0.91, the 100-year 6-h areal rainfall at the location = 0.91 (89) = 81 mm or 13.5 mm/h for 6 h.

It is necessary here to differentiate between 24-h rainfall and 1-day rainfall. Comparing M5 values for stations recording hourly data and those recording only in days, the equivalent M5 depths for 1, 2, 4 and 8 *rainfall days* were found. To convert rainfall days to rainfall hours it is necessary to multiply the M5 values for days by the factors given in table 2.10.

2.9.5 Storm profile. The FSR describes a series of storm profiles of varying probability for both winter and summer. For specific proposals, several, of varying peakedness, should be used in the application of 'design rain'. However, for the purposes of this text, and for use in flood estimation in later chapters, only the 'winter 75 per cent profile' will be described. This is the storm profile that, on average, is more peaked than 75 per cent of all winter storm profiles; see figure 2.17 and table 2.11. Rain is assumed to fall at varying intensity, but symmetrically about the peak intensity.

TABLE 2.10 *Factors to give M5 values for hours from rainfall days* [13]

Rainfall days	1	2	4	8
Multiplying factor	1.11	1.06	1.03	1.015
Rainfall hours	24	48	96	192

Table 2.10 is used subsequently in the calculation of RSMD (see example 7.2).

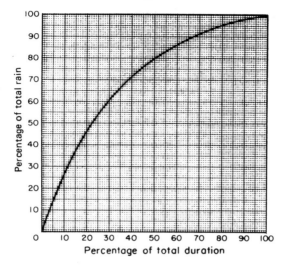

Figure 2.17 *The 'winter 75 per cent' storm profile* [13]

TABLE 2.11 *The 'winter 75 per cent' storm profile*

Cumulative duration (% of total)	Cumulative rain (% of total)	Incremental % rain in % time		Incremental intensity/ Mean intensity
5	13.5	13.5	5	2.7
10	26	12.5	5	2.5
20	46	20	10	2.0
30	60	14	10	1.4
40	72	12	10	1.2
50	80	8	10	0.8
60	86	6	10	0.6
70	91	5	10	0.5
80	95	4	10	0.4
90	98	3	10	0.3
100	100	2	10	0.2

The profile is usually used in the form of a stepped profile as shown in figure 7.30 (page 187). When used in this way it is convenient, though not essential, to divide the rain duration into an uneven number of parts, to permit symmetrical arrangement about a peak intensity. (See example 7.2 [15] on page 186.)

2.10 Probable maximum precipitation (PMP)

The preceding section describes methods for estimating rainfall for particular locations with prescribed frequencies. If one imagines lower and lower frequencies then the rainfall amounts apparently would continue to increase. That, however, is a statistical concept rather than a physical one and begs the question – is there some maximum possible value, limited by the nature of the earth's atmosphere and the laws of physics? This uppermost limit, assuming it exists, is known as probable maximum precipitation (PMP) and is defined [16] as the theoretically greatest depth of precipitation for a given duration that is physically possible over a given size of storm area for a particular location and time of year.

   This concept of an upper limit to possible precipitation was known previously (before 1950) as maximum possible precipitation but PMP is now preferred because, being an estimate, it has a degree of uncertainty.

2.10.1 The physical method. The essential requirement for precipitation is a supply of moist air. The air's water content is measured by its temperature and dew-point, which are standard meteorological observations and are therefore often available in catchments where few flow records exist. The amount of water in the atmosphere above a catchment can be established by the assumption of a saturated adiabatic lapse rate (see section 2.3) or by actual weather balloon observations. The amount is usually between 10 mm and 60 mm in depth. This water is precipitated by cooling, which is accomplished almost always by vertical movement; the air mass expands adiabatically thus precipitating moisture, which in turn liberates latent heat and so the vertical movement is accelerated. The process is self-stimulated and in extreme cases is responsible for the 'cloud-burst' of the severe thunder-storm. The vertical lifting may also be due to orographic or frontal effects as discussed in section 2.6.

   The inflow of moist air at the base of an ascending storm column can be measured by wind-speed observations around storm peripheries. These observations yield data about the amount of water that may be carried into a storm column and hence precipitated. The data can also be inferred from synoptic weather charts if these are based on a sufficiently dense network of observer stations.

   There is an established relationship between this convergence of air masses, their subsequent vertical motion and the condensation of the water vapour within them. So any study of PMP starts from data on observed areal-averaged rainfall. Extreme rainfalls in a region give guidance to maximum convergence rates and vertical air speeds which in turn produce maximum rates of condensation and precipitation. The procedures for determining PMP involve maximising observed storm rainfall by factoring up the actual water in an observed event to the limiting value that the air at its temperature could have carried; by transposing storms that have occurred in the region to the actual catchment considered, and by ensuring that *envelopment* is used. This last term means the use of enveloping curves to all observed data of depth, area and duration used in the derivation of study values. Detailed descriptions of the methods are given in WMO's *Operational Hydrology Report No. 1* [17].

Maxima for separate seasons or months can be subsequently derived by taking maxima of recorded seasonal or monthly values of air moisture and air inflow and assuming that they occur simultaneously. These can be compared with historic storms that have occurred in the catchment and meteorologically-similar neighbouring regions. In this way an array of data about the historic storms can be built up and compared with the hypothetical maxima.

The PMP can be derived from the hypothetical maxima by taking the peak of the envelope curve covering them. It is argued sometimes that the elements used to define the maxima should themselves be subject to frequency analysis before the adoption of a PMP. This, however, presupposes that the physical processes of rain-producing storms are random events that can occur in vastly different magnitudes. This may not be true to the same extent as it is of precipitation recorded at a rain gauge or the flood in a river. Also, in taking the known maximum values of the determining factors and combining them in space over the catchment, the concept of return period is invalidated. The approach is deterministic rather than statistical.

PMP varies with geographical location, season of the year, area and elevation of the catchment and storm duration.

2.10.2 The statistical method.

An alternative statistical method is sometimes used when sufficient precipitation data are available. It is particularly useful when other meteorological information about dew points, wind speeds, etc., is unavailable. It is also much quicker to perform.

The procedure due to Hershfield [18, 19] is based on the general frequency equation [20]

$$X_t = \bar{X}_n + K\,\sigma_n$$

where X_t is the rainfall of a specified duration for return period t, and $\bar{X}_n$ and σ_n are, respectively, the mean and standard deviation for a series of annual maximum values of rainfall of that duration.

Then, if X_m is the maximum observed rainfall

$$X_m = \bar{X}_n + K_m\,\sigma_n$$

where K_m is the number of standard deviations that must be added to the mean to find the maximum.

A study in the USA [18] used records of 24 h rainfall for over 2700 stations with minimum periods of record of 10y. $\bar{X}_n$ and σ_n were calculated conventionally but each station's maximum recorded rain was omitted. The largest value of K_m found to satisfy the omitted maxima was 15.

Values of K_m for durations other than 24 h were established [19] and are as illustrated in figure 2.18.

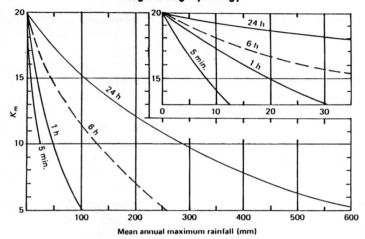

Figure 2.18 K_m *as a function of rainfall duration and mean of annual series. (Reprinted from Journal American Water Works Association, Vol. 57, No. 8 (August 1965) by permission. Copyright © 1965 American Water Works Association)*

It will be noted that this figure has a maximum K_m value of 20, but other investigators have found even higher values in other regions [21]. K_m apparently varies inversely with intensity so the greater the annual rainfall in a particular climate the less likely that K_m will be greater than 15. The more arid the area, the greater the likelihood.

Detailed descriptions of the methods are given in WMO's *Operational Hydrology Report No. 1* [17].

2.10.3 General application of data. Many large storms have been analysed, particularly in the USA, and maximum values of rainfall depth for various durations and areas have been published [16, 21]. Such data are usually presented as sets of curves, each representing a rainfall depth plotted on rectangular co-ordinates of storm duration in hours and storm area in square miles. These are a useful guide to limiting values but must be used with judgement for particular catchments, since topography and elevation as well as climate may modify the result appreciably for other regions.

In the light of this section it is interesting to look again at the world maximum point rainfalls quoted in section 2.8.2 where it is suggested that if rainfall is plotted against duration, both scales being logarithmic, the world's greatest observed point rainfalls lie on or just under a straight line whose equation is

$$R = 16.6D^{0.475}$$

where R is a rainfall in inches and D is duration in hours. Reference to figure 2.4 indicates that the extreme point rainfalls recorded in the British Isles, extracted from more than half a million station years of data, lie close to a line of equation

$$R = 106D^{0.46}$$

where R is in mm and D in hours. The fit is particularly good in the region from 0.2 to 20 hours, which is probably the critical range for application to unit hydrographs.

References

1. MAIDENS, A. L. New Meteorological Office rain-gauges. *Meteorological Magazine*, **94**, No. 1114 (May 1965) 142
2. GOODISON, C. E. and BIRD, L. G. Telephone interrogation of rain-gauges. *Meteorological Magazine*, **94**, No. 114 (May 1965) 144
3. GREEN, M. J. Effects of exposure on the catch of rain gauges. *Technical Publication 67, Water Research Association*, July 1969
4. BLEASDALE, A. Rain gauge networks development and design with special reference to the United Kingdom. *International Association of Scientific Hydrology Symposium on Design of Hydrological Networks, Quebec,* 1965
5. PAULHUS, J. L. H. Indian ocean and Taiwan rainfalls set new records. *Monthly Weather Rev.*, **93** (May 1965) 331
6. BILHAM, E. G. *The Classification of Heavy Falls of Rain in Short Periods*, H.M.S.O., London, 1962 (republished)
7. A guide for engineers to the design of storm-sewer systems. *Road Research Laboratory, Road Note 35*, H.M.S.O., London, 1963
8. HOLLAND, D. J. Rain intensity–frequency relationships in Britain. *British Rainfall 1961*, H.M.S.O., London, 1967
9. YARNALL, D. L. Rainfall intensity–frequency data. U.S. *Department of Agriculture Miscellaneous Publication*, **204**, Washington D.C., 1935
10. LINSLEY, R. K. and KOHLER, M. A. Variations in storm rainfall over small areas. *Trans. Am, Geophys. Union*, **32** (April 1951) 245
11. HOLLAND, D. J. The Cardington rainfall experiment. *Meteorological Magazine*, **96**, No. 1140 (July 1967) 193–202
12. YOUNG, C. P. Estimated rainfall for drainage calculations. *LR 595, Road Research Laboratory*, H.M.S.O., London, 1973
13. Natural Environmental Research Council. *Flood Studies Report*, Vols. I–V, NERC, 1975
14. The areal reduction factor in rainfall frequency estimation. *FSR Suppl. Report No. 1*, Institute of Hydrology, Wallingford, United Kingdom, 1977
15. THIESSEN, A. H. Precipitation for large areas. *Monthly Weather Rev.*, **39**, (July 1911) 1082
16. HANSEN, E. M., SCHREINER, L. C. and MILLER, J. F. Application of probable maximum precipitation estimates – United States east of the 105th meridian. *Hydrometeorological Report No. 52*, National Weather Service, National Oceanic and Atmospheric Administration, US Department of Commerce, Washington D.C., 1982
17. Manual for estimation of probable maximum precipitation. *Operational Hydrology Report No. 1*, 2nd edn, World Meteorological Organisation, 1986
18. HERSHFIELD, D. M. Estimating the probable maximum precipitation. *Proc. Am. Soc. Civ. Eng., Journal Hydraulics Division*, **87** (1961) 99

19. HERSHFIELD, D. M. Method for estimating probable maximum precipitation. *Journal American Waterworks Association*, 57 (1965) 965
20. CHOW, V. T. A general formula for hydrologic frequency analysis. *Trans. Am. Geophys. Union*, 32 (1961) 231
21. MCKAY, G. A. *Statistical estimates of precipitation extremes for the Prairie Provinces*, Canada Department of Agriculture, PFRA Engineering Branch, 1965

Further reading

BINNIE, G. M. and MANSELL-MOULLIN, M. The estimated probable maximum storm and flood on the Jhelum River—a tributary of the Indus. *Symposium on River Flood Hydrology, Inst. Civ. Eng., London*, 1966, Paper No. 9

GLASSPOOLE, J. Heavy falls in short periods (two hours or less). *Quart. J. Roy. Meteorological Soc.*, 58 (1931) 57–70

Guide to hydrometeorological practices. *U.N. World Meteorological Org. No. 168, Technical Publication 82*, United Nations, Geneva, 1965

Handbook of Meteorology (ed. by Berry, Bollay and Beers), McGraw-Hill, New York, 1949, p. 1024

HERSCHFIELD, D. and WILSON, W. T. Generalising of rainfall intensity–frequency data. *Proc. Int. Assoc. Sci. Hydrol., General Assembly of Toronto*, 1 (1957) 499–506

JENNINGS, A. H. World's greatest observed point rainfalls. *Monthly Weather Rev.*, 78, (January 1950) 4

LANGBEIN, W. B. Hydrologic data networks and methods of extrapolating or extending available hydrologic data. *Flood Control Series No. 15*, United Nations, 1960

PARTHASARATHY, K. and GURBACHAN SINGH. Rainfall intensity–duration-frequencies for India, for local drainage design. *Indian J. Meteorology Geophys.*, 12 (1961) 231–42

PETERSON, K. R. A precipitable water nomogram. *Bull. Am Meterological Soc.*, 42 (1961) 199

SOLOT, S. Computation of depth of precipitable water in a column of air. *Monthly Weather Rev.*, 67 (1939) 100

Standards for methods and records of hydrologic measurements. *Flood Control Series No. 6*, United Nations, 1954

Problems

2.1 An air mass is at a temperature of 28 °C with relative humidity of 70 per cent. Determine: (a) saturation vapour pressure, (b) saturation deficit, (c) actual vapour pressure in mbar and mm Hg, (d) dew-point, and (e) wet-bulb temperature.

2.2 Discuss the relationships between depth, duration and area of rainfall for particular storms.

2.3 The following are annual rainfall figures for four stations in Derbyshire. The average values for Cubley and Biggin School have not been established.

	Average (in.)	1959	1960
Wirksworth	35·5	26·8	48·6
Cubley		19·5	42·4
Rodsley	31·3	21·6	42·1
Biggin School		33·1	54·2

(a) Assume departures from normal are the same for all stations. Forecast the Rodsley 'annual average' from that at Wirksworth over the two years of record. Compare the result with the established value.

(b) Forecast annual averages for Cubley and Biggin School using both Wirksworth and Rodsley data.

(c) Comment on the assumption in part (a). Is it reasonable?

2.4 One of four monthly-read rain gauges on a catchment area develops a fault in a month when the other three gauges record 37, 43 and 51 mm respectively. If the average annual precipitation amounts of these three gauges are 726, 752 and 840 mm respectively and of the broken gauge 694 mm, estimate the missing monthly precipitation at the latter.

2.5 Compute the average annual rainfall, in inches depth, on the catchment area shown
(i) by arithmetic means, (ii) by the Theissen method, and (iii) by plotting isohyets. Comment on the applicability of each method.

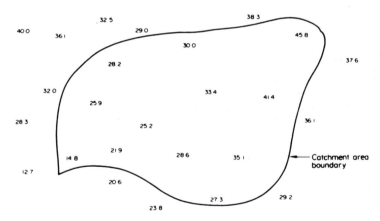

2.6 Discuss the setting of rain gauges on the ground and comment on the effect of wind and rain falling non-vertically on the catch.

2.7 Annual precipitation at rain gauge X and the average annual precipitation at twenty surrounding rain gauges are listed in the following table

Year	Annual precipitation (mm)		Year	Annual precipitation (mm)	
	Gauge X	20-station average		Gauge X	20-station average
1972	188	264	1954	223	360
1971	185	228	1953	173	234
1970	310	386	1952	282	333
1969	295	297	1951	218	236
1968	208	284	1950	246	251
1967	287	350	1949	284	284
1966	183	236	1948	493	361
1965	304	371	1947	320	282
1964	228	234	1946	274	252
1963	216	290	1945	322	274
1962	224	282	1944	437	302
1961	203	246	1943	389	350
1960	284	264	1942	305	228
1959	295	332	1941	320	312
1958	206	231	1940	328	284
1957	269	234	1939	308	315
1956	241	231	1938	302	280
1955	284	312	1937	414	343

(a) Examine the consistency of station X data.

(b) When did a change in regime occur? Discuss possible causes.

(c) Adjust the data and determine what difference this makes to the 36-year annual average precipitation at station X.

2.8 Plot the data for the mean of the 20 stations in 2.7 as a time series. Then plot 5-year moving averages and accumulated annual departures from the 36-year mean. Is there evidence of cyclicity or particular trends?

2.9 At a given site, a long-term wind-speed record is available for measurements at heights of 10 m and 15 m above the ground. For certain calculations of evaporation the speed at 2 m is required, so it is desired to extend the long-term record to the 2 m level. For one set of data the speeds at 10 m and 15 m were 9.14 and 9.66 m/s respectively.

(a) What is the value of the exponent relating the two speeds and elevations?

(b) What speed would you predict for the 2 m level?

2.10 A rainfall gauge registers a fall of 9 mm in 10 minutes.

(a) How frequently would you expect such a fall at a particular place in Britain?

(b) What total volume of rain would be expected to fall on 3 km² surrounding the gauge?

2.11 What is the maximum 1-day rainfall expected in Britain for a 50-year period at location X (average annual rainfall 1000 mm) and a 30-year period at location Y (average annual rainfall 1750 mm)?

2.12 What is the average rainfall over an area of 8 km^2 during a storm lasting 30 minutes with a frequency of once in 20 years in (a) Oxford, (b) Kumasi. Does your answer for (b) require qualification?

2.13 The table below lists the annual maximum rainfall over a 6-hour period, for an observation station, for 20 years. Make a first estimate of the Probable Maximum Precipitation of 6-hour duration at the station. Is this station in the UK?

Depths in mm

146	193
203	124
194	176
102	210
173	102
151	131
182	165
115	187
143	121
180	98

3 Evaporation and Transpiration

3.1 Meteorological factors

Evaporation is important in all water resource studies. It affects the yield of river basins, the necessary capacity of reservoirs, the size of pumping plant, the consumptive use of water by crops and the yield of underground supplies, to name but a few of the parameters affected by it.

Water will evaporate from land, either bare soil or soil covered with vegetation, and also from trees, impervious surfaces like roofs and roads, open water and flowing streams. The rate of evaporation varies with the colour and reflective properties of the surface (the *albedo*) and is different for surfaces directly exposed to, or shaded from, solar radiation.

In moist temperate climates the loss of water through evaporation is typically 600 mm per year from open water and perhaps 450 mm per year from land surfaces. In an arid climate, like that of Iraq, the corresponding figures could be 2000 mm and 100 mm, the great disparity in this latter case being caused by absence of precipitation for much of the year.

Some of the more important meteorological factors affecting evaporation are discussed below.

Solar radiation. Evaporation is the conversion of water into water vapour. It is a process that is taking place almost without interruption during the hours of daylight and often during the night also. Since the change of state of the molecules of water from liquid to gas requires an energy input (known as the latent heat of vaporisation), the process is most active under the direct radiation of the sun. It follows that clouds, which prevent the full spectrum of the sun's radiation reaching the earth's surface, will reduce the energy input and so slow up the process of evaporation.

Wind. As the water vaporises into the atmosphere, the boundary layer between earth and air, or sea and air, becomes saturated and this layer must be removed and continually replaced by drier air if evaporation is to proceed. This movement of the air in the boundary layer depends on wind and so is a function of wind speed.

Relative humidity. The third factor affecting evaporation is the relative humidity of the air. As the air's humidity rises, its ability to absorb more water vapour decreases and the rate of evaporation slows. Replacement of the boundary layer of saturated air by air of equally high humidity will not maintain the evaporation rate: this will occur only if the incoming air is drier than the air that is displaced.

Temperature. As mentioned above, an energy input is necessary for evaporation to proceed. It follows that if the ambient temperatures of the air and ground are high, evaporation will proceed more rapidly than if they are low, since heat energy is more readily available. Since the capacity of air to absorb water vapour increases as its temperature rises, so air temperature has a double effect on how much evaporation takes place, while ground and water temperatures have single direct effects.

3.2 Transpiration

Growing vegetation of all kinds needs water to sustain life, though different species have very different needs. Only a small fraction of the water needed by a plant is retained in the plant structure. Most of it passes through the roots to the stem or trunk and is *transpired* into the atmosphere through the leafy part of the plant.

In field conditions it is practically impossible to differentiate between evaporation and transpiration if the ground is covered with vegetation. The two processes are commonly linked together and referred to as *evapotranspiration.*

The amount of moisture that a land area loses by evapotranspiration depends primarily on the incidence of precipitation, secondly on the climatic factors of temperature, humidity etc. and thirdly on the type, manner of cultivation and extent of vegetation. The amount may be increased, for example, by large trees whose roots penetrate deeply into the soil, bringing up and transpiring water that would otherwise be far beyond the influence of surface evaporation.

Transpiration proceeds almost entirely by day under the influence of solar radiation. At night the pores or *stomata* of plants close up and very little moisture leaves the plant surfaces. Evaporation, on the other hand, continues so long as a heat input is available, although it occurs primarily during the day. The other factor of importance is the availability of a plentiful water supply. If water is always available in abundance for the plant to use in transpiration, more will be used than if at times less is available than could be used. Accordingly, a distinction must be made between *potential evapotranspiration* and what actually takes place. Most of the methods of estimation necessarily assume an abundant water supply and so give the potential figure.

3.3 Methods of estimating evaporation

3.3.1 Water budget or storage equation approach. This method consists of drawing up a balance sheet of all the water entering and leaving a particular catchment or drainage basin. If rainfall is measured over the whole area on a regular and systematic basis then a close approximation to the amount of water arriving from the atmosphere can be made. Regular stream gauging of the streams draining the area, and accurately prepared flow-rating curves, will indicate the water leaving the area by surface routes. The difference between these two can be accounted for in only three ways:

(i) by a change in the storage within the catchment, either in surface lakes and depressions or in underground aquifers;
(ii) by a difference in the underground flow into and out of the catchment;
(iii) by evaporation and transpiration.

The storage equation can be written generally as

$$E = P + I \pm U - O \pm S$$

where E = evapotranspiration
 P = total precipitation
 I = surface inflow (if any)
 U = underground outflow
 O = surface outflow
 S = change in storage (both surface and subsurface).

If the observations are made over a sufficiently long time the significance of S, which is not cumulative, will decrease and can be ignored if the starting and finishing points of the study are chosen to coincide as nearly as possible with the same seasonal conditions. The significance of U cannot be generalised but in many cases can be assigned second-order importance because of known geological conditions that preclude large underground flows. In such cases a good estimation of evapotranspiration becomes possible and the method provides a means of arriving at first approximations.

3.3.2 Energy budget method. This method, like the water budget approach, involves solving an equation that lists all the sources and sinks of thermal energy and leaves evaporation as the only unknown. It involves a great deal of instrumentation and is still under active development. It cannot be used readily without many data that are not normally available, and so it is a specialist approach.

3.3.3 Empirical formulae. Many attempts have been made to produce satisfactory formulae for the estimation of evaporation. These are usually for evaporation from an open water surface, as indeed are the more general methods to follow. The reason for this is simple. Evaporation, if it is to take place, presupposes a supply of water. Whatever the meteorological conditions may be, if there is no water present then there can be no evaporation. Accordingly, estimating methods using meteorological data work on the assumption that abundant water is available; that is, a free water surface exists. The results obtained therefore are not necessarily a measure of actual but of *potential evaporation*. Often these two are the same, as for example, in reservoirs where a free water surface exists. When evaporation from land surfaces is concerned, the loss of water in this way clearly depends on availability: rainfall, water-table level, crop or vegetation, and soil type all have an influence, which can be expressed by applying an empirical factor, usually less than unity, to the free water surface evaporation.

There are two cases that should be considered:

(i) when the temperature of the water surface is the same as the air temperature;
(ii) when the air and water surface temperatures are different.

Case (i) rarely occurs and is empirically treated by the equation

$$E_a = C(e_s - e) f(u) \qquad (3.1)$$

where E_a = open water evaporation per unit time (for air and water temperature the same $t\,^\circ C$) in mm/day
C = an empirical constant
e_s = saturation vapour pressure of the air at $t\,^\circ C$ (mm mercury)
e = actual vapour pressure in the air above (mm mercury)
u = wind speed at some standard height (m).

The following equation has been empirically obtained for this case and is of general validity

$$E_a = 0.35(e_s - e)(0.5 + 0.54u_2) \qquad (3.2)$$

u_2 denotes wind speed in m/s at a height of 2 m: E_a is in mm/day.

Case (ii) is the one that normally occurs. Again a formula should have the form

$$E_o = C(e'_s - e) f(u) \qquad (3.3)$$

but now e'_s is the saturation vapour pressure of the boundary layer of air between air and water, whose temperature t'_s is not the same as either air or water and is virtually impossible to measure. Accordingly empirical formulae have been developed in the form of equation 3.1, which work fairly well for specific locations where the constants have been derived, but have no general validity.

Such a formula, derived for the Ijsselmeer in The Netherlands, and *only applicable to it and similar conditions*, is

$$E_o = 0.345(e_w - e)(1 + 0.25u_6)$$

where E_o = evaporation of the lake in mm per day
e_w = saturation vapour pressure at temperature t_w of the surface water of the lake in mm mercury

e = actual vapour pressure in mm mercury

u_6 = wind velocity in m/s at a height of 6 m above the surface.

3.3.4 Penman's Theory. The following nomenclature is used:

E_o = evaporation from open water (or its equivalent in heat energy)

e_w = saturation vapour pressure of air at water surface temperature t_w

e = actual vapour pressure of air at temperature t = saturation vapour pressure at dew-point t_d

e_s = saturation vapour pressure of air at temperature t

e_s' = saturation vapour pressure of air at boundary layer temperature t_s'

n/D = cloudiness ratio = actual/possible hours of sunshine

R_A = Angot's value of solar radiation arriving at the atmosphere

R_C = sun and sky radiation actually received at earth's surface on a clear day

R_I = net amount of radiation absorbed at surface after reflection

R_B = radiation *from* the earth's surface.

In 1948 Penman [1] presented a theory and formula for the estimation of evaporation from weather data. The theory is based on two requirements, which must be met if continuous evaporation is to occur. These are: (i) there must be a supply of energy to provide latent heat of vaporisation; (ii) there must be some mechanism for removing the vapour, once produced.

The energy supply. During the hours of daylight there is a certain measurable amount of short-wave radiation arriving at the earth's surface. The amount depends on latitude, season of the year, time of day and degree of cloudiness. Assuming there were no clouds and a perfectly transparent atmosphere, the total radiation to be expected at a point has been given in tabular form by Angot, and is reproduced in table 3.1 as values of R_A.

If R_C = short-wave radiation actually received at the earth from sun and sky and n/D = ratio of actual/possible hours of sunshine, then Penman gives (for southern England)

$$R_C = R_A(0.18 + 0.55n/D)$$

and quotes Kimball (for Virginia, USA)

$$R_C = R_A(0.22 + 0.54n/D)$$

and Prescott (for Canberra, Australia)

$$R_C = R_A(0.25 + 0.54n/D)$$

Thus even on days of complete cloud cover (n/D = 0), about 20 per cent of solar radiation reaches the earth's surface, while on cloudless days about 75 per cent of radiation gets through.

Part of R_C is reflected as short-wave radiation; the exact amount depends on the reflectivity of the surface, or, the reflection coefficient r.

TABLE 3.1 Angot's values of short-wave radiation flux R_A at the outer limit of the atmosphere in $g\ cal/cm^2/day$ as a function of the month of the year and the latitude[a]

Latitude (degrees)	Jan.	Feb.	Mar.	Apr.	May	June	July	Aug.	Sept.	Oct.	Nov.	Dec.	Year
N 90	0	0	55	518	903	1077	944	605	136	0	0	0	3540
80	0	3	143	518	875	1060	930	600	219	17	0	0	3660
60	86	234	424	687	866	983	892	714	494	258	113	55	4850
40	358	538	663	847	930	1001	941	843	719	528	397	318	6750
20	631	795	821	914	912	947	912	887	856	740	666	599	8070
Equator	844	963	878	876	803	803	792	820	891	866	873	829	8540
20	970	1020	832	737	608	580	588	680	820	892	986	978	8070
40	998	963	686	515	358	308	333	453	648	817	994	1033	6750
60	947	802	459	240	95	50	77	187	403	648	920	1013	4850
80	981	649	181	9	0	0	0	0	113	459	917	1094	3660
S 90	995	656	92	0	0	0	0	0	30	447	932	1110	3540

[a]From: *Physical and Dynamical Meteorology* by David Brunt, p. 112 (Cambridge University Press, 1944). The SI unit for R_A is joules/m^2/day. The table in g cal/cm^2/day is used so that it is compatible with Rijkoort's nomogram. The conversion is 1 g cal/cm^2 = 41.9 kJ/m^2.

If R_I = the net amount of radiation absorbed, then (for southern England)

$$R_I = R_C(1 - r) = R_A(1 - r)(0.18 + 0.55n/D)$$

In turn, some of R_I is re-radiated by the earth as long-wave radiation, particularly at night when the air is dry and the sky clear. The net outward flow R_B may be expressed empirically as

$$R_B = \sigma T_a^4 (0.47 - 0.077\sqrt{e})(0.20 + 0.80n/D)$$

where σ = Lummer and Pringsheim constant = 117.74×10^{-9} g cal/cm^2/day
$\quad\quad T_a$ = absolute earth temperature = $t\,^\circ C + 273$
$\quad\quad e$ = actual vapour pressure of air in mm mercury.

Hence the net amount of energy finally remaining at a free water surface ($r = 0.06$) is given by H, where

$$\begin{aligned}
H &= R_I - R_B \\
&= R_C - rR_C - R_B \\
&= R_C(1 - r) - R_B \\
&= R_A(0.18 + 0.55n/D)(1 - 0.06) - R_B
\end{aligned}$$

Therefore

$$H = R_A(0.18 + 0.55n/D)(1 - 0.06) - (117.4 \times 10^{-9})$$
$$T_a^4(0.47 - 0.077\sqrt{e})(0.20 + 0.80n/D) \quad\quad\quad (3.4)$$

This heat is used up in four ways, that is

$$H = E_o + K + S + C$$

where E_o = heat available for evaporation from open water
$\quad\quad K$ = convective heat transfer from the surface
$\quad\quad S$ = increase in heat of the water mass (that is, storage)
$\quad\quad C$ = increase in heat of the environment (negative advected heat).

Over a period of days and frequently over a single day, the storage of heat is small compared with the other changes, and the same is true of environmental storage, so that to a small degree of error

$$H = E_o + K$$

Vapour removal. It has been shown that evaporation may be represented by

$$E_o = C(e_s' - e)f(u)$$

but that e_s' cannot be evaluated if air and water are at different temperatures. Penman now made the assumption that the transport of vapour and the transport of heat by eddy diffusion are essentially controlled by the same mechanism (that is, atmospheric turbulence), the one being governed by $(e_s' - e)$, the other by $(t_s' - t)$. To a close approximation therefore

$$\frac{K}{E_o} = \beta = \frac{\gamma(t'_s - t)}{e'_s - e}$$

where γ = psychrometer constant
= 0.66 if t is in $^\circ$C and e in mbar

Now since

$$H = E_o + K = E_o(1 + \beta)$$

then

$$E_o = \frac{H}{1 + \beta} = \frac{H}{1 + \gamma \dfrac{t'_s - t}{e'_s - e}}$$

Now eliminate $t'_s - t$ by substitution, since $t'_s - t = (e'_s - e_s)/\Delta$

where e_s = saturation vapour pressure at temperature t
Δ = slope of vapour pressure curve at t, = $\tan \alpha$ (see figure 3.1).

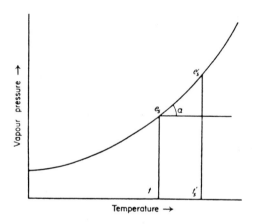

Figure 3.1 *Saturation vapour pressure curve*

This is reasonable since t'_s is never very far from t.
Hence

$$E_o = \frac{H}{1 + \dfrac{\gamma}{\Delta} \cdot \dfrac{e'_s - e_s}{e'_s - e}} \tag{3.5}$$

Now e_s must be eliminated.

Since

$$e'_s - e_s = (e'_s - e) - (e_s - e) \tag{3.6}$$

and from equation 3.1

$$E_a = C(e_s - e) f(u)$$

while from equation 3.3

$$E_o = C(e_s' - e) f(u)$$

then

$$\frac{E_a}{E_o} = \frac{e_s - e}{e_s' - e} \tag{3.7}$$

where E_a = evaporation (in energy terms) for the hypothetical case of equal temperatures of air and water.

Then by the values of equations 3.6 and 3.7 into equation 3.5

$$E_o = \frac{H}{1 + \dfrac{\gamma}{\Delta}\left[\dfrac{(e_s' - e) - (e_s - e)}{e_s' - e}\right]}$$

and

$$E_o = \frac{H}{1 + \dfrac{\gamma}{\Delta}\left(1 - \dfrac{E_a}{E_o}\right)}$$

from which

$$E_o = \frac{\Delta H + \gamma E_a}{\Delta + \gamma}$$

Δ has values obtained from the saturation vapour pressure curve, typically as shown

$t =$		$\Delta =$
0 °C		0.36
10		0.61
20		1.07
30		1.80

Referring to equations 3.2 and 3.4 for E_a and H respectively, it can be seen that E_o is now computed from standard meteorological observations of mean air temperature, relative humidity, wind velocity at a standard height and hours of sunshine. The formula has been checked in many parts of the world and gives very good results. Being based on physical principles it is of general application and gives values that should serve for most project studies until supplemented by actual evaporation measurements (see section 3.6).

To overcome the computational labour involved in solving the Penman equation, a nomogram has been designed by P. J. Rijkoort of the Royal Meteorological Institute, The Netherlands, which enables rapid evaluations to be made.

It is reproduced as appendix C at the back of this book by permission of the designer. The nomogram has been drawn for a slightly different value of R_C from that used by Penman

$$R_C = R_A(0.20 + 0.48n/D)$$

instead of Penman's $R_A(0.18 + 0.55n/D)$, but any difference will be smaller than the probable margin of error in cloud cover estimation, so it can be ignored.

Values of R_A can be derived for any latitude from table 3.1.

Where actual sunshine records are available, then table 3.2 may be used to determine n/D accurately. When sunshine records are not available, n/D may be estimated from assessment of cloud cover in tenths; i.e. ten-tenths cloud is completely overcast and 0 tenths completely clear. Unity minus this fraction may be used for n/D. For example, on a day when about seven-tenths of the sky is covered by cloud, on average, then

$$n/D = (1 - 7/10) = 0.3$$

Several observations per day should be made.

3.4 Evaporation from land surfaces using Penman's E_o value

Penman subsequently reported on results of experiments conducted on turfed soil and bare soil to determine how their evaporation rates (E_T and E_B) compared with open water (E_o) [2]. He concluded that the evaporation rate from a freshly wetted bare soil was about 90 per cent of that from an open water surface exposed to the same weather. That is

$$E_B/E_o = 0.90$$

For grassed surfaces the comparison was more erratic, and provisional figures for turf with a plentiful water supply were given as follows:

Values of E_T/E_o for southern England

November to February	0.6
March to April September to October	0.7
May to August	0.8
Whole year	0.75

These figures are all less than unity because of the greater reflectivity of vegetation compared to open water and also because the transpiration of plants virtually ceases at night.

3.5 Thornthwaite's formulae for evapotranspiration

Thornthwaite carried out many experiments in the USA using lysimeters and extensively studied the correlation between temperature and evapotranspiration. From this work [3] he devised a method enabling estimates to be made of the

TABLE 3.2 Mean daily maximum hours of sunshine for different months and latitudes[a]

Latitude North South	Jan. July	Feb. Aug.	March Sept.	April Oct.	May Nov.	June Dec.	July Jan.	Aug. Feb.	Sept. March	Oct. April	Nov. May	Dec. June
50°	8.5	10.1	11.8	13.8	15.4	16.3	15.9	14.5	12.7	10.8	9.1	8.1
48°	8.8	10.2	11.8	13.6	15.2	16.0	15.6	14.3	12.6	10.9	9.3	8.3
46°	9.1	10.4	11.9	13.5	14.9	15.7	15.4	14.2	12.6	10.9	9.5	8.7
44°	9.3	10.5	11.9	13.4	14.7	15.4	15.2	14.0	12.6	11.0	9.7	8.9
42°	9.4	10.6	11.9	13.4	14.6	15.2	14.9	13.9	12.5	11.1	9.8	9.1
40°	9.6	10.7	11.9	13.3	14.4	15.0	14.7	13.7	12.5	11.2	10.0	9.3
35°	10.1	11.0	11.9	13.1	14.0	14.5	14.3	13.5	12.4	11.3	10.3	9.8
30°	10.4	11.1	12.0	12.9	13.6	14.0	13.9	13.2	12.4	11.5	10.6	10.2
25°	10.7	11.3	12.0	12.7	13.3	13.7	13.5	13.0	12.3	11.6	10.9	10.6
20°	11.0	11.5	12.0	12.6	13.1	13.3	13.2	12.8	12.3	11.7	11.2	10.9
15°	11.3	11.6	12.0	12.5	12.8	13.0	12.9	12.6	12.2	11.8	11.4	11.2
10°	11.6	11.8	12.0	12.3	12.6	12.7	12.6	12.4	12.1	11.8	11.6	11.5
5°	11.8	11.9	12.0	12.2	12.3	12.4	12.3	12.3	12.1	12.0	11.9	11.8
0°	12.1	12.1	12.1	12.1	12.1	12.1	12.1	12.1	12.1	12.1	12.1	12.1

[a]From: Crop Water Requirements. Irrigation and Drainage Paper 24 (United Nations F.A.O., Rome, 1975).

potential evapotranspiration from short, close-set vegetation with an adequate water supply, in the latitudes of the USA.

If t_n = average monthly temperature of the consecutive months of the year in °C (where n = 1, 2, 3, . . . , 12) and j = monthly 'heat index', then

$$j = \left(\frac{t_n}{5}\right)^{1.514} \tag{3.8}$$

and the yearly 'heat index', J, is given by

$$J = \sum_{1}^{12} j \text{ (for the 12 months)}$$

The potential evapotranspiration for any month with average temperature t (°C) is then given, as PE_x, by

$$PE_x = 16 \left(\frac{10t}{J}\right)^a \text{ mm per month}$$

where

$$a = (675 \times 10^{-9})J^3 - (771 \times 10^{-7})J^2 + (179 \times 10^{-4})J + 0.492 \tag{3.9}$$

However PE_x is a theoretical standard monthly value based on 30 days and 12 hours of sunshine per day. The actual PE for the particular month with average temperature t (°C) is given by

$$PE = PE_x \frac{DT}{360} \text{ mm} \tag{3.10}$$

where D = number of days in the month

T = average number of hours between sunrise and sunset in the month.

The method has been tested by Serra, who suggested that equations (3.8) and (3.9) may be simplified as follows

$$j = 0.09t_n^{3/2}$$

$$a = 0.016J + 0.5$$

This method of estimating potential evapotranspiration is empirical and complicated and requires the use of a nomogram for its solution. Thornthwaite published such a nomogram, which is reproduced in figure 3.2.

The first step is to obtain the heat index J. From figure 3.2 obtain the unadjusted value of potential evapotranspiration by drawing a straight line from the location's J value through the point of convergence at t = 26.5 °C. (If t is greater than 26.5 °C, use the table alongside figure 3.2.) PE_x for the month can then be read off, corresponding to its given mean temperature. Twelve values are obtained for each of the 12 months. These unadjusted values can then be adjusted for day

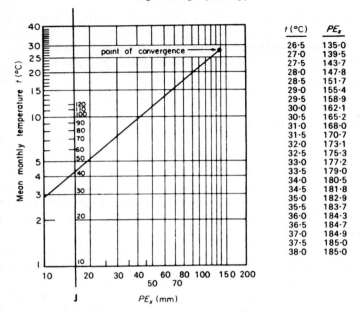

Figure 3.2 *Nomogram and table for finding potential evapotrans-piration PE_x (after C. W. Thornthwaite. Courtesy of 'Geographical Review'; copyright American Geographic Society, New York)*

and month length by equation 3.10 and totalled to give annual potential evapo-transpiration.

It has been found that the method gives reasonably good results whatever the vegetation cover, though different types of vegetation will affect a particular locality's true value. The formula is based on temperature, which does not necessarily correspond to incoming solar radiation immediately, because of the 'heat inertia' of land and water. Transpiration, however, responds directly to solar radiation. Accordingly, care should be exercised when using the method to ensure that conditions do not change abruptly in a particular month, though if figures for many consecutive months are being used, the cumulative differences are probably negligible.

For project studies the method is a useful complement to the Penman approach, though it will be found generally to give higher values of potential evaporation than the latter.

3.6 Direct measurement of evaporation by pans

Whenever possible, direct observations of evaporation should be made. The instrument used for this is the evaporation pan. In Britain the standard pan is 1.83 m (6 ft.) square and 610 mm (2 ft.) deep filled to a depth of 550 mm

(1 ft 9 in.) and set in the ground so that the rim of the pan projects 76 mm (3 in.) above the surrounding ground. Regular observations of pan evaporation are made at some 30 points throughout the country [4].

In the USA the standard or Class A pan is circular 1.22 m (4 ft.) in diameter and 254 mm (10 in.) deep, filled to a depth of 180 mm (7 in.) set on a timber grillage with the pan bottom 150 mm (6 in.) above ground level. Regular evaporation readings are taken at over 400 places.

A third type of pan is sometimes used in the United Kingdom, the Peirera pan, which is circular like the Class A pan but deeper and sunk in the ground with a 3 in. air space surrounding it.

The water levels in evaporation pans are prescribed and have to be set daily after measurement. Measurements are usually by hook-gauge, with allowance made for input rainfall. Because monthly and yearly totals are the sums of many small differences, each with the same chance of error, the observations must be made with extreme care.

The Class A pan has a greater daily range of temperature than the square one, but is usually homogeneous whereas the water in the square pan may stratify. Doubling the wind run may increase evaporation by up to 20 per cent.

The relatively small capacities and shallow depths of pans in comparison to lake and river volumes and their situation at or near the land surface allows proportionately greater amounts of advected heat from the atmosphere to be absorbed by the water in the pan through the sides and bottom, than by natural open water, and by some pans more than others. Pan evaporation is therefore usually too high and a pan coefficient has to be applied. These coefficients range from 0.65 to greater than unity, depending on the dimensions and siting of the pan. Generally the standard British pan has a coefficient about 0.92 and the U.S. Weather Bureau Class A pan about 0.75 but there are quite wide variations. Law [5] carried out comparative tests over 14 years at two sites in Yorkshire and found the ratio of evaporation from the Class A pan to that from the square British pan ranged between 1.17 and 1.40 with an average of 1.32. Houk [6] gives a full account of known American values and Olivier [7] quotes many data from African and Near-Eastern sources.

There are difficulties in using pans for the direct measurement of evaporation, arising from the difficulty of measuring very small differences of elevation and the subsequent application of coefficients to relate the measurements from a small tank to large bodies of open water. Nevertheless, actual field measurements should form an important part of any project studies of evaporation.

3.7 Consumptive use

Evapotranspiration is the term used for the evaporation of moisture from the earth's surface including lakes and streams and the vegetation that may cover the land. *Consumptive use* refers to the evaporation and transpiration from vegetation-covered land areas only, frequently with respect to horticulture and agriculture and associated irrigation requirements. The terms are often used synonymously.

The consumptive use of water in an area is dependent on many factors, including climate, the supply of soil moisture, growing vegetation, type of soil and methods of land management. Climatic factors include precipitation, temperature, humidity, wind and latitude of the locality (which affects the length of the growing season). Soil moisture supply depends on topography and underground flow, as well as precipitation. Soil types and land management vary widely over short distances. There are no formulae of general validity but several empirical formulae can be used with local coefficients to determine annual water use in any locality within certain broad limitations.

3.7.1 Arable crops. Consumptive use refers to water that is actually used, while evapotranspiration formulae give potential water use. Reference to figure 3.3 will show that for a particular locality, unless the rain falls when it is needed, a large water deficit may develop in the growing season despite quite high rainfall. In a case like this, consumptive use will be less than potential evapotranspiration and a need for irrigation in the growing period is indicated.

Modern methods of estimation of consumptive use include a development of the Penman open-water evaporation calculation and a more empirical method developed from the Blaney–Criddle procedures of the Division of Irrigation and

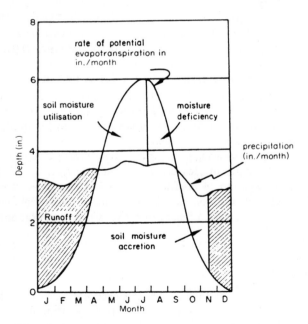

Figure 3.3 *Typical annual soil moisture deficiency diagram* [8]

Water Conservation, Soil Conservation Service of the US Dept of Agriculture [9].

Both these and other methods are fully set out with worked examples in the relevant UN Food and Agriculture Organisation publication [10].

Since both methods require reference to many figures and tables, it is not feasible to provide comprehensive accounts in this text. Nevertheless, summaries of the two approaches are given below.

3.7.2 Modified Penman method. This method depends on meteorological measurements and is most likely to give satisfactory estimates of crop water requirements.

The original Penman equation yielded evaporation from an open-water surface, E_0, through consideration of the two terms H and E_a covering energy (radiation) and aerodynamics (wind and humidity).

The E_0 term was modified to yield evapotranspiration from a grassed land surface by multiplying by a crop coefficient of about 0.8. This has been shown to be a reasonable approximation where calm conditions apply, not only in England where it was first derived, but in tropical and semi-arid regions as well. However, in windier climates the crop coefficient varies widely from 0.8.

A new standard or reference crop evapotranspiration ET_0 is introduced. This is defined as

"the rate of evapotranspiration from an extended surface of 8 to 15 cm tall green grass cover of uniform height, actively growing, completely shading the ground and not short of water" [10]

The value of ET_0 for a particular climate and location is calculated from

$$ET_0^* = W \times R_n + (1 - W) \times f(u) \times (e_s - e)$$

where ET_0^* is the unadjusted reference crop evapotranspiration in mm/day
  W = a temperature-related weighting factor
  R = a net radiation in equivalent evaporation in mm/day
  $f(u)$ = wind-related function
  $(e_s - e)$ = the difference between saturated vapour pressure at mean air temperature and mean actual vapour pressure of the air in mb.

ET_0 is then found from ET_0^* through adjustment for day and night-time weather conditions. The correlation is given graphically for several conditions of varying wind and humidity.

Having determined a value of ET_0, the crop evapotranspiration $ET_{(crop)}$, i.e. the crop's water requirement is found from

$$ET_{(crop)} = k_c \times ET_0$$

where k_c is a crop coefficient, representing the evapotranspiration of a particular crop grown under optimum conditions (for the climate and location) and producing optimum yields.

Values of k_c for extensive lists of different crops, grown under different conditions of wind and humidity, and at various stages of growth, are available [10]. Table 3.3 is a representative sample of the k_c values for different crops.

TABLE 3.3   *Typical crop coefficient k_c for crops at different growth stages and prevailing climatic conditions[a]*

Crop	Humidity	Wind (m/s)	Min. relative humidity > 70%		Min. relative humidity < 20%	
			0-5	5-8	0-5	5-8
Barley	mid-season	3	1.05	1.1	1.15	1.2
	harvest	4	0.25	0.25	0.2	0.2
Carrots		3	1.0	1.05	1.1	1.15
		4	0.7	0.75	0.8	0.85
Cotton		3	1.05	1.15	1.2	1.25
		4	0.65	0.65	0.65	0.7
Cucumber		3	0.9	0.9	0.95	1.0
		4	0.7	0.7	0.75	0.8
Flax		3	1.0	1.05	1.1	1.15
		4	0.25	0.25	0.2	0.2
Melons		3	0.95	0.95	1.0	1.05
		4	0.65	0.65	0.75	0.75
Groundnuts		3	0.95	1.0	1.05	1.1
		4	0.55	0.55	0.6	0.6
Potatoes		3	1.05	1.1	1.15	1.2
		4	0.7	0.7	0.75	0.75
Soyabeans		3	1.0	1.05	1.1	1.15
		4	0.45	0.45	0.45	0.45
Tomatoes		3	1.05	1.1	1.2	1.25
		4	0.6	0.6	0.65	0.65
Wheat		3	1.05	1.1	1.15	1.2
		4	0.25	0.25	0.2	0.2

[a] After: *Crop Water Requirements. Irrigation and Drainage Paper 24* (United Nations F.A.O., Rome, 1975).

3.7.3 Modified Blaney–Criddle method.   The original method, developed by the US Dept of Agriculture [9] involves temperature t and percentage of daylight hours p, to derive a monthly consumptive use factor f. Relative humidity is not considered.

f is then multiplied by k, a monthly consumptive use coefficient for a particular crop to obtain monthly consumptive use in inches.

In the modified method, the monthly consumptive use factor f is expressed in mm and degrees Celsius ($^\circ$C) as

$$f = p(0.46t + 8.13)$$

where t is the mean of daily maximum and minimum temperature over the month considered and p is the mean daily percentage of annual daytime hours. The factor f is then expressed in mm/day representing the mean value over the given month.

From these f values, graphical relationships giving equivalent ET_0 values are provided for three levels of relative humidity, three levels of wind speed and three levels of cloudiness (n/D).

Figure 3.4 is a composite representation of three of the nine figures in [10].

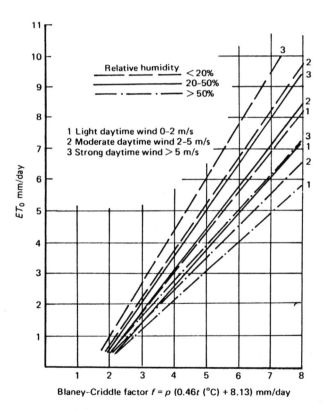

Figure 3.4 ET_0 against f for medium cloudiness (n/D: 0.6–0.8), 3 wind conditions and 3 relative humidities. After Crop Water Requirements: Irr. and Drainage Paper 24 (United Nations F.A.O., Rome, 1975)

Table 3.4 gives the mean daily percentage (p) of annual daylight hours for different latitudes.

The ET_0 and k_c values for both the modified Penman and Blaney–Criddle methods are compatible, so only one set of k_c values is required.

3.7.4 Forest. The first important work on afforestation of reservoired catchments was published by Law [11]. He subsequently observed a lysimeter of 450 m² of Sitka spruce forest for a period of 14 years. Some of his observations were published in earlier editions of this text. In summary Law pointed out that forested upland catchments resulted in greater water losses than if the cover was grass. This was not generally accepted at first but Law's work led to the setting in train of much more extensive long-term investigations by the Institute of Hydrology. These have been conducted over 15 years on several catchments in the United Kingdom, of which the best known and most intensively investigated are the Plynlimon catchments of the Rivers Severn and Wye. These studies have been reported at intervals during the period [12, 13] and have led to valuable and sometimes surprising conclusions. Briefly, these are as follows.

(1) With annual rainfall around 2300 mm, there was a considerably greater loss from the forested catchment than from the grassland. The forested catchment losses were about double those of the grassland — about 850 mm against 405 mm on average.

(2) The reason for the heavier forest loss is the interception of rainfall by the forest canopy and the subsequent evaporation from it.

(3) In drier parts of the country, where the rainfall was about 600 mm annually, forested catchments were no more likely to lead to losses than grassland, and in some observations of pine forest were actually less [14].

(4) The Penman formula for evapotranspiration is in good agreement with the actual figure for grassland catchments but underestimates the figure considerably for forested catchments in wetter parts of the country and during wet periods in areas of low rainfall.

(5) Although erosion protection has always been an aim of afforestation, it is not always achieved. Where open ditching has been adopted to drain forest areas, the erosion may be greater than for grassland. This was the case in the Plynlimon catchments, where sediment was measured over five years.

(6) From observations of lysimeters, it has been found that about 50 to 55 per cent of rainfall at the forest canopy falls through to the forest floor, while a further 12 to 23 per cent reaches ground level by stem-flow; so that between 22 and 38 per cent evaporates from the canopy.

(7) Mature forest cover reduces peak flows by about one-third compared with grassland, but where drainage ditches have been dug for tree planting there may well be both an increase compared with grassland and shorter response times, particularly while the trees are immature.

TABLE 3.4 Mean daily percentage (p) of annual daytime hours for different latitudes [10]

| Latitude North | Jan. | Feb. | Mar. | Apr. | May | June | July | Aug. | Sept. | Oct. | Nov. | Dec. |
South	July	Aug.	Sept.	Oct.	Nov.	Dec.	Jan.	Feb.	Mar.	Apr.	May	June
60°	0.15	0.20	0.26	0.32	0.38	0.41	0.40	0.34	0.28	0.22	0.17	0.13
58°	0.16	0.21	0.26	0.32	0.37	0.40	0.39	0.34	0.28	0.23	0.18	0.15
56°	0.17	0.21	0.26	0.32	0.36	0.39	0.38	0.33	0.28	0.23	0.18	0.16
54°	0.18	0.22	0.26	0.31	0.36	0.38	0.37	0.33	0.28	0.23	0.19	0.17
52°	0.19	0.22	0.27	0.31	0.35	0.37	0.36	0.33	0.28	0.24	0.20	0.17
50°	0.19	0.23	0.27	0.31	0.34	0.36	0.35	0.32	0.28	0.24	0.20	0.18
48°	0.20	0.23	0.27	0.31	0.34	0.36	0.35	0.32	0.28	0.24	0.21	0.19
46°	0.20	0.23	0.27	0.30	0.34	0.35	0.34	0.32	0.28	0.24	0.21	0.20
44°	0.21	0.24	0.27	0.30	0.33	0.35	0.34	0.31	0.28	0.25	0.22	0.20
42°	0.21	0.24	0.27	0.30	0.33	0.34	0.33	0.31	0.28	0.25	0.22	0.21
40°	0.22	0.24	0.27	0.30	0.32	0.34	0.33	0.31	0.28	0.25	0.22	0.21
35°	0.23	0.25	0.27	0.29	0.31	0.32	0.32	0.30	0.28	0.25	0.23	0.22
30°	0.24	0.25	0.27	0.29	0.31	0.32	0.31	0.30	0.28	0.26	0.24	0.23
25°	0.24	0.26	0.27	0.29	0.30	0.31	0.31	0.29	0.28	0.26	0.25	0.24
20°	0.25	0.26	0.27	0.28	0.29	0.30	0.30	0.29	0.28	0.26	0.25	0.25
15°	0.26	0.26	0.27	0.28	0.29	0.29	0.29	0.28	0.28	0.27	0.26	0.25
10°	0.26	0.27	0.27	0.28	0.28	0.29	0.29	0.28	0.28	0.27	0.26	0.26
5°	0.27	0.27	0.27	0.28	0.28	0.28	0.28	0.28	0.28	0.27	0.27	0.27
0°	0.27	0.27	0.27	0.27	0.27	0.27	0.27	0.27	0.27	0.27	0.27	0.27

Water engineers concerned with land use in reservoir catchments should be aware that if annual rainfall is higher than average then afforestation will increase water loss, whatever its other benefits.

References

1. PENMAN, H. L. Natural evaporation from open water, bare soil and grass. *Proc. Roy. Soc.*, A 193 (April 1948) 120
2. PENMAN, H. L. Evaporation over the British Isles. *Quart. J. Roy. Met. Soc.*, 76 (1950) 372
3. THORNTHWAITE, C. W. An approach towards a rational classification of climate. *Geographical Rev.*, 38 (1948) 55
4. *British Rainfall 1939* (and subsequent years), H.M.S.O., London
5. LAW, F. The aims of the catchment studies at Stocks Reservoir, Slaidburn, Yorkshire. Unpublished communication to Pennines Hydrological Group, Institution of Civil Engineers, September 1970
6. HOUK, I. E. *Irrigation Engineering*, Vol. 1, Wiley, New York, 1951
7. OLIVIER, H. *Irrigation and Climate*, Arnold, London, 1961
8. THORNTHWAITE, C. W. The moisture factor in climate. *Trans. Am. Soc. Civ. Eng.*, 27, No. 1 (February 1946) 41
9. BLANEY, H. F. and CRIDDLE, W. D. Determining water requirements in irrigated areas from climatological and irrigation data. *Div. Irr. Water Conserv., S.C.S. U.S. Dept. Agr.*, SCS-TP-96, Washington D.C., 1950
10. DOORENBOS, J. and PRUITT, W. O. Crop water requirements. *Irrigation and Drainage Paper 24. F.A.O.*, United Nations, Rome, 1975
11. LAW, F. The effect of afforestation upon the yield of water catchment areas. *J. Brit. Waterworks Assoc.*, 38 (1956) 484
12. Water balance of the headwater catchments of the Wye and Severn 1970–75. *Report No. 33. Institute of Hydrology*, Wallingford, United Kingdom, December 1976
13. NEWSON, M. D. The results of ten years' experimental study on Plynlimon, Mid-Wales, and their importance for the water industry. *J. Inst. Water Eng. Sci.*, 33 (1979) 321–33
14. GASH, J. H. C. and STEWART, J. B. The evaporation from Thetford Forest during 1975. *J. Hydrol.*, 35 (1977) 385–96

Further reading

BLANEY, H. F. Definitions, methods and research data. A symposium on the consumptive use of water. *Trans. Am. Soc. Civ. Eng.*, 117 (1952) 949

BLANEY, H. F. and CRIDDLE, W. D. Determining consumptive use and irrigation water requirements. *USDA (ARS) Tech. Bull. 1275*, 1962

CRIDDLE, W. D. Consumptive use of water and irrigation requirements. *J. Soil Water Conserv.*, 1953

CRIDDLE, W. D. Methods of computing consumptive use of water. Paper 1507. *Proc. Am. Soc. Civ. Eng.*, 84 (January 1958)

FAO/UNESCO. *International Sourcebook on Irrigation and Drainage of Arid Lands*, UNESCO, Paris, Hutchinson, London, 1973

FORTIER, S. and YOUNG, A. A. *Bull. U.S. Dept. Agr. 1340* (1925), *185* (1930), *200* (1930), *379* (1933)

HARRIS, F. S. The duty of water in Cache Valley, Utah. *Utah Agr. Exp. Sta. Bull.*, **173**, 1920

HICKOX, G. H. Evaporation from a free water surface. Paper 2266. *Trans. Am. Soc. Civ. Eng.*, **111** (1946), and discussion by C. Rohwer.

HILL, R. A. Operation and maintenance of irrigation systems. Paper 2980. *Trans. Am. Soc. Civ. Eng.*, **117** (1952) 77

HORSFALL, R. A. Planning irrigation projects. *J. Inst. Eng. Australia*, 22, No. 6 (June 1950)

LOWRY, R. and JOHNSON, A. R. Consumptive use of water for agriculture. Paper 2158. *Trans. Am. Soc. Civ. Eng.*, **107** (1942), and discussion by R. E. Rule, E. E. Foster, H. F. Blaney and R. W. Davenport

Measurement and Estimation of Evaporation and Evapotranspiration. *Technical Note 83*, World Meteorological Organisation, 1966

PENMAN, H. L. Estimating evaporation. *Trans. Am. Geophys. Union*, 31 (February 1956) 43

ROHWER, C. Evaporation from different types of pans. *Trans. Am. Soc. Civ. Eng.*, **99** (1934) 673

STEWART, J. I. and HAGAN, R. M. Functions to predict effects of crop water deficits. *ASCE J. Irrigation and Drainage Div.*, 99 (1973) 421

STEWART, J. J., HAGAN, R. M. and PRUITT, W. O. Function to predict optimal irrigation programs. *ASCE, J. Irrigation and Drainage Div.*, 100 (1974) 179

WHITE, W. N. A method of estimating ground water supplies based on discharge by plants and evaporation from soil. Results of investigations in Escalante Valley, Utah. *U.S. Geological Survey Water Supply Paper 659-A*, Washington D.C., 1932

Problems

3.1 Determine the evaporation from a free water surface using the Penman equation nomogram for the following cases

Locality	Month	Temp.	h	n/D	U_2
Amsterdam (52°N)	July	18°C	0.5	0.5	1.2 m/s
Seattle (47°N)	Jan.	4°C	0.8	0.3	1.5 m/s
Khartoum (16°N)	June	30°C	0.2	1.0	0.9 m/s

3.2 Use the nomogram for the solution of Penman's equation to predict the daily potential evapotranspiration from a field crop at latitude 40°N in April, under the following conditions: mean air temperature = 20°C; mean $h = 70$ per cent; sky cover = 60 per cent cloud; mean U_2 = 2.5 m/s; ratio of potential evapotranspiration to potential evaporation = 0.7.

3.3 Compute the potential evapotranspiration according to Thornthwaite for two locations A and B where the local climate yields the following data

	A	B	Daylight hours of year at A (per cent)		A	B	Daylight hours of year at A (per cent)
Jan.	−5	−2	6	July	19	16	11
Feb.	0	2	7	Aug	17	14	10
Mar.	5	3	7½	Sept.	13	10	8½
Apr.	9	7	8½	Oct.	9	8	7½
May	13	10	10	Nov.	5	3	7
June	17	15	11	Dec.	0	0	6

(a) at A for April (mean temperature 10°C) and November (mean temperature 3°C),

(b) at B for June (mean temperature 20°C) and October (mean temperature 8°C).

At A the average number of hours between sunrise and sunset is 13 for April and 9 for November. At B the figures are 14 for June and 10 for October. Use the Serra simplification for A and the nomogram for B.

3.4 Water has maximum density at 4°C; above and below this temperature its density is less. Consider a deep lake in a region where the air temperature falls below 4°C in winter.

(a) Describe what will happen in the lake in spring and autumn.

(b) What will be the effect of what happens on (i) the time lag between air and water temperatures? (ii) the evaporation rate in the various seasons?

(c) Will there be a difference if the winter temperature does not drop below 4°C and if so explain why.

3.5 Using the modified Blaney–Criddle method, determine the crop water requirement for growing cotton in mid-season in very dry conditions at a location at latitude 30°N in February where the mean daily temperature is 28°C, and the mean wind speed is 4 m/s.

3.6 Discuss the advantages and disadvantages of evaporation pans placed above the ground surface (for example, the U.S. Class A pan) compared to those sunk in the earth.

3.7 Draw up a water-budget for 100 units of rainfall falling on to a coniferous forest in a temperate coastal climate. Describe the processes involved and indicate the proportions of the rain that becomes involved in each.

3.8 Describe fully Penman's evaporation theory for open water surfaces. Show how each parameter used affects the evaporation and discuss how the theory differs from other evaporation formulae.

3.9 A large reservoir is located in latitude 40°30'N. Compute monthly and annual lake evaporation for the reservoir from the given data using the nomogram of Penman's theory. If the Class A pan evaporation at the reservoir for the year is 1143 mm, compute the pan coefficient. Assuming that the precipitation on the lake is as given and that the runoff represents unavoidable spillage of this precipitation during floods, what is the net annual anticipated loss from the reservoir per square kilometre of surface in cubic metres per day?

What would the change in *evaporation* be for the month of July if the reservoir was at 40°S?

	Mean air temp. (°C)	Dew-point (°C)	Average wind speed (m/s)	Cloud coverage in tenths	Precipitation (mm)	Runoff (mm)
Oct.	14.4	7.8	0.8	5.9	51	—
Nov.	8.3	1.7	1.3	7.2	99	23
Dec.	3.9	2.2	1.7	9.5	102	43
Jan.	2.2	1.9	2.1	8.7	117	58
Feb.	2.2	1.4	2.2	6.3	91	20
Mar.	4.4	1.1	1.3	5.1	69	12
Apr.	8.9	3.3	1.1	3.4	51	—
May	15	10	0.9	2.6	28	—
June	20	15.6	0.8	0.2	3	—
July	23.9	16.7	0.75	0.1	0	—
Aug.	22.8	17.8	0.7	0.0	0	—
Sept.	17.8	12.8	0.75	1.5	20	8

3.10 It is proposed to develop a lake in a mountainous region for water supply. It is at latitude 20°N. June appears to be the critical month and mean June values of various parameters are as follows:

air temperature 24°C;   relative humidity 0.6;
wind speed 2 m above surface, 2.5 m/s;   cloud cover 10%

The lake has a surface area of 300 km^2. A river flows into the lake and its long-average inflow for June is 28 m^3/s.
Calculate the net inflow to the reservoir for an average June.

4 Infiltration and Percolation

4.1 Infiltration capacity of soil

When rain falls upon the ground it first of all wets the vegetation or the bare soil. When the surface cover is completely wet, subsequent rain must either penetrate the surface layers if the surface is permeable, or run off the surface towards a stream channel if the surface is impermeable.

If the surface layers are porous and have minute passages available for the passage of water droplets, the water *infiltrates* into the subsurface soil. Soil with vegetation growing on it is always permeable to some degree. Once infiltrating water has passed through the surface layers, it *percolates* downwards under the influence of gravity until it reaches the zone of saturation at the *phreatic surface*.

Different types of soil allow water to infiltrate at different rates. Each soil type has a different infiltration capacity, f, measured in mm/h or in./h. For example, it can be imagined that rain falling on a gravelly or sandy soil will rapidly infiltrate and, provided the phreatic surface is below the ground surface, even heavy rain will not produce surface runoff. Similarly a clayey soil will resist infiltration and the surface will become covered with water even in light rains. The rainfall rate, i, also obviously affects how much rain will infiltrate and how much will run off.

4.2 Factors influencing f_c

Nassif and Wilson [1] carried out extensive studies on infiltration using a weighable laboratory catchment 25 m² in area. They concluded that for any soil under constant rainfall, infiltration rate decreases in accordance with an equation of the form first used by Horton [2].

$$f = f_c + \mu e^{-Kt}$$

where f = infiltration rate at any time t (mm/h)
f_c = infiltration capacity at large value of t (mm/h)

66

$\mu = f_0 - f_c$

f_0 = initial infiltration capacity at $t = 0$ (mm/h)

t = time from beginning of rainfall (min)

K = constant for a particular soil and surface (min^{-1}).

K is a function of surface texture: if vegetation is present K is small, while a smoother surface texture, such as bare soil, will yield larger values.

f_0 and f_c are functions of both soil type and cover. For example, a bare sandy or gravelly soil will have high values of f_0 and f_c and a bare clayey soil will have low values of f_0 and f_c, but both values will increase for both soils if they are turfed.

f_c is a function of: (i) slope up to a limiting value (varying between 16 and 24 per cent), after which there is little variation, (ii) initial moisture content—the drier the soil initially, the larger will be f_c, but the variation may be comparatively small, and (iii) rainfall intensity—if the intensity i increases, f_c increases, and this parameter has a greater effect on f_c than any other variable.

These correlations are illustrated in figure 4.1 for a typical agricultural soil. Table 4.1 lists some representative values of K, f_0 and f_c for different soil types. The parameters K and f_0 are relatively stable for particular soils and do not vary noticeably with slope of catchment or rain intensity, f_c on the other hand, varies widely with both and so is shown as a range of values.

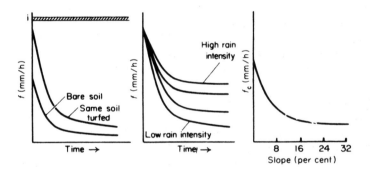

Figure 4.1   *The variation of infiltration capacity*

TABLE 4.1   *Representative values of K, f_0 and f_c for different soil types*

Soil type		f_0 (mm/h)	f_c (mm/h)	K (min^{-1})
Standard	bare	280	6–220	1.6
agricultural	turfed	900	20–290	0.8
Peat		325	2–20	1.8
Fine sandy	bare	210	2–25	2.0
clay	turfed	670	10–30	1.4

Until recently it had been generally thought that f_c was a constant for a particular soil, but this does not now seem to be so. The infiltration rate appears to be largely controlled by the surface pores. Even quite a small increase in the hydrostatic head over these pores results in an increase in the flow through the soil surface. If the surface layer is imagined as shown in figure 4.2 where the

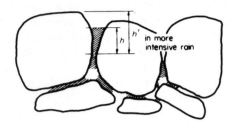

Figure 4.2 *Hydraulic head on soil-grain pores*

surface soil grains are shown, then the governing factor is the head h over the smallest cross-section of a pore. This continues to increase with rainfall intensity until a limiting value is reached where runoff prevents any further increase. It seems unlikely that this limiting condition is often reached in natural conditions.

Previous researchers [3] have found similar results but attributed the increase in f_c at higher rainfall intensities to lack of homogeneity in their experimental catchment watershed. Others have also emphasised the overriding importance of the *superficial layer* of a soil [4].

The infiltration rate of a soil is the sum of percolation and water entering storage above the groundwater table. Generally the soil is far from saturated and so storage goes on increasing for very long periods. Accordingly, f_c goes on decreasing under a steady rain intensity for equally long periods.

However the soil will eventually become saturated under persistent rain. All the reservoirs of interstitial space become filled. When precipitation ceases the soil gradually loses the 'free' water, to a point where it can sustain the water content against gravity, i.e. water will no longer drain from it. At this point the soil is said to be at *field capacity*.

Exposed soils can be rendered almost impermeable by the compacting impact of large drops coupled with the tendency to wash very fine particles into the voids. The surface tends to become 'puddled' and the f_c value drops sharply. Similarly, compaction due to man or animals treading the surface, or to vehicular traffic, can severely reduce infiltration capacity.

Dense vegetal cover such as grass or forest tends to promote high values of f_c. The dense root systems, all providing ingress to the subsoil, the layer of organic debris forming a sponge-like surface, burrowing animals and insects opening up ways into the soil, the cover preventing compaction and the vegetation's transpiration removing soil moisture, all tend to help the infiltration process.

Other effects that marginally affect the issue are frost heave, leaching out of soluble salts and drying cracks which increase f_0, and the entrapping of interstitital air, which decreases f_0. Temperature has some effect since flow in interstices is laminar and hence viscosity has a direct effect on resistance to flow. Other things being equal, f_0 and f_c will have higher values during the warmer seasons of the year.

4.3 Methods of determining infiltration capacity

4.3.1 Infiltrometers.
An infiltrometer is a wide-diameter, short tube, or other impervious boundary surrounding an area of soil. Usually two such rings are used concentrically as shown in figure 4.3. The rings are flooded to a depth of

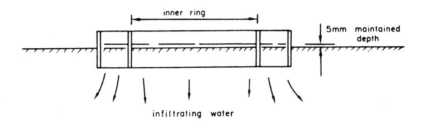

Figure 4.3   *Infiltrometer*

5 mm over the surface and continually refilled to maintain this depth, the inflow to the central tube being measured. The purpose of the outer tube is to eliminate to some extent the edge effect of the surrounding drier soil. Such tests give useful comparative results but they do not simulate real conditions and have been largely replaced by sprinkler tests on larger areas. Here the sprinkler simulates rainfall, and runoff from the plot is collected and measured as well as inflow, the difference being assumed to have infiltrated.

While rain-simulating sprinklers are a good deal more realistic than flooded rings, there are limitations to the reliability of results thus obtained, which usually give higher values of f than natural conditions do. For qualitative effects (for example, comparisons between different conditions of vegetation, soil types etc) the methods are simple and effective.

Consistent and repeatable results can be obtained by using laboratory catchments with rainfall simulators, where the quantity and thickness of soil is adequately representative of nature. Nassif and Wilson [1] used 7 tonnes of soil in a layer 200 mm thick, and measured all inputs, outputs and changes of storage. Such equipment as theirs gives very good comparative and perhaps absolute figures for infiltration, but it still does not simulate natural conditions completely as there is atmospheric pressure at the bottom of the laboratory soil layer, while this is not so in nature.

4.3.2 Drainage basin rainfall–runoff analysis. Several investigators have attempted to improve on sprinkled infiltrometers by choosing small 'homogeneous' drainage basins and carefully measuring precipitation, evaporation and outflow as surface runoff. By eliminating everything except infiltration, average f values can be obtained for such basins by techniques presented by Horton [5] and Sherman [6].

The difficulty remains of ensuring that there has been neither unrecorded underground outflow nor variation in underground aquifer storage, so that although quantitative results are obtained the analysis is intricate and the margin for error is wide.

4.3.3 Φ-index method. In practice, it is possible to obtain *infiltration indices*, which enable reasonable approximations to be made of infiltration losses. One of these is the *Φ-index*, which is defined as the average rainfall intensity above which the volume of rainfall equals the volume of runoff. In figure 4.4 a rain-

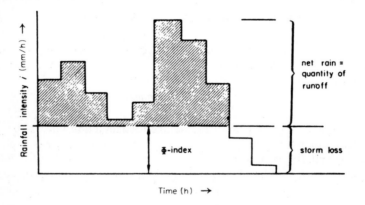

Figure 4.4   *Infiltration loss by Φ-index*

storm is shown plotted on a time base in terms of its average hourly intensity. The shaded area above the dashed line represents measured runoff, as mm, over the catchment area. Since the unshaded area below the line is measured rainfall but did not appear as runoff, it represents all the losses, including surface detention and evaporation as well as infiltration. However, infiltration is much the largest loss in many catchments and although it is a rough and ready method (since it takes no account of the variation in f with time) it is widely used as a means of quickly assessing probable runoff from large catchments for particular rainstorms.

Example 4.1.   Given a total rainfall of 75 mm as shown in figure 4.5(a) and a surface runoff equivalent to 33 mm, establish the Φ-index for the catchment.

The Φ-index line is drawn so that the shaded area above it contributes 33 mm of runoff. In this case the index is 8 mm/h.

Suppose, however, the same total rainfall had been distributed as shown in figure 4.5(b). To obtain a runoff of 33 mm above the Φ-index line requires the

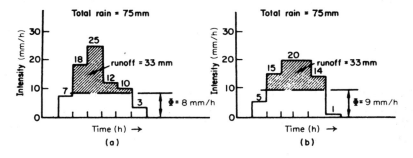

Figure 4.5 *Examples of Φ-index computations*

line to be raised to give an index value of 9 mm/h. It is seen therefore that one determination of the Φ-index is of limited value and that many such determinations should be made, and averaged, before the index is used.

4.3.4 The f_{av} method. This method is a developed version of the Φ-index in that it attempts to allow for depression storage and short rainless periods during a storm, as well as eliminating all rain periods where the rainfall intensity is less than the infiltration capacity assumed.

Referring to figure 4.6, the approximate position of the f_{av} line is fixed by reference to runoff data and the raingraph. The line is then moved vertically until the various losses are balanced and the runoff values satisfied. The loss

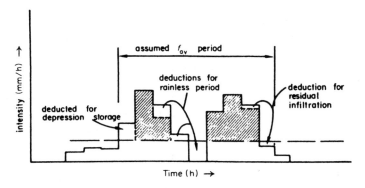

Figure 4.6 *Example of f_{av} distribution*

estimates are based on whatever data are available and on the judgment of the
analyst and hence are obviously subjective but, from the analysis of many
storms, values of f_{av} can be deduced for various conditions.

In applying a derived value of f_{av} to a rainstorm to predict expected runoff,
the rain periods lying outside the f_{av} period are assumed lost and the net rain is
found directly after inserting the estimated losses as shown. Butler [7] gives a
detailed account of the method.

4.4 Soil moisture

4.4.1 Initial soil-moisture conditions. The foregoing methods of estimating
losses are essentially based on records of rainfall and runoff for particular
catchments and their behaviour under rains of varying intensity. They show
average limiting values of infiltration capacity, obtaining this information in
terms of the whole catchment rather than by sampling very small areas as the
infiltrometer does. They do not, however, enable predictions to be made with
any accuracy of the amount of rainfall that will be absorbed by the soil and so
lost to runoff in a particular case, since this depends, among other factors, on
the state of wetness of the soil at the start of rainfall. As well as affecting runoff
through storage capacity, the initial soil moisture affects the infiltration capacity
and hence the runoff in the initial stages of a storm. Some other measurement of
this parameter is therefore necessary if forecasts of runoff are to be made from
stipulated rainfalls. Φ-index or f_{av} methods provide only average values, which
may, in a particular case, be far removed from actuality, and these methods are
best used after a separate assessment of initial losses.

There are two approaches to this problem discussed here. The first is the
antecedent precipitation index, used in the USA, and the second, the *estimated
soil-moisture deficit*, used in Britain.

4.4.2 Antecedent precipitation index. The antecedent precipitation index is
based on the premise that soil moisture is depleted at a rate proportional to the
amount in storage in the soil. There is therefore a logarithmic relationship

$$I_t = I_o k^t \tag{4.1}$$

where I_0 = initial value of index (mm)
$\quad\quad I_t$ = index value t days later
$\quad\quad k$ = a recession constant with a value about 0.92 but varying between
$\quad\quad\quad$ 0.85 and 0.98.

If t is unity, then any day's value is k times that of the previous day. If precipitation occurs it will increase the value of the index by an amount that is indeterminate, since some rain may have left the catchment as surface runoff. The

amount added to the index should therefore strictly be the *basin recharge* only, but the difference in the index by using all the precipitation is usually small.

The progressive daily reduction of the index is due to evapotranspiration, which alters seasonally, so equation 4.1 is used with a k value that also varies seasonally. When the index is used to assess the runoff that takes place from a particular rainstorm, this variation can be incorporated into a graphical coaxial relationship derived from the analysis of a larger number of observed rainfall and runoff data on a particular catchment. Linsley *et al.* [8] give a detailed description of how such relationships can be derived, and figure 4.7 is reproduced to illustrate the technique.

The diagram is entered at the antecedent precipitation index and a horizontal line is followed until the particular week-number curve corresponding to the calendar date is met. From the intersection a line is dropped vertically down to intersect with the line of appropriate storm duration in hours, and then followed

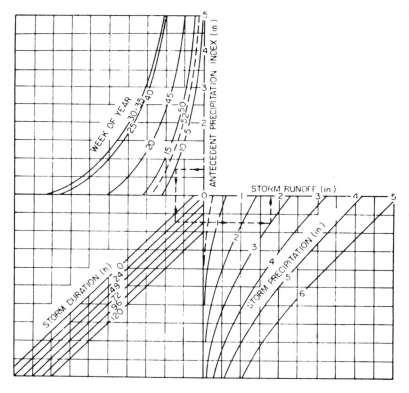

Figure 4.7 *Storm–runoff relationship for Monacacy River at Jug Bridge, Maryland, USA. (After U.S. Weather Bureau. From 'Hydrology for Engineers' by R. K. Linsley, M. A. Kohler and J. L. H. Paulhus. McGraw-Hill Book Co., New York, 1958)*

horizontally across to intersect with the line of total storm precipitation. A vertical line from the last intersection indicates the appropriate runoff.

The antecedent precipitation index is a valuable tool in obtaining probable runoff forecasts for well-documented catchments but much work is necessary to derive relationships like figure 4.7.

4.4.3 Estimated soil-moisture deficit. Since evapotranspiration is continually removing the soil moisture and precipitation is replacing it, continuous regular measurements of the parameters influencing these two processes provide a means of estimating the soil-moisture deficit (or *s.m.d.*) for any location at any time. This avoids the use of a single recession constant, such as k in equation 4.1.

When evapotranspiration exceeds precipitation, vegetation draws on accumulated soil-moisture to continue transpiration. As *s.m.d.* increases, the vegetation begins to exhibit signs of stress and eventually begins to wilt. The determination of *s.m.d.* is therefore particularly important for assessing irrigation needs and for hydrological studies of water supply.

In Britain the UK Meteorological Office has developed a sophisticated method for the estimation of weekly and monthly evaporation and soil-moisture deficits over the whole country — the Meteorological Office Rainfall and Evaporation Calculation System: MORECS [9]. It has five main components:

(i) Data collection, interpolation and averaging.
(ii) Data analysis to obtain evaporative demand (i.e. potential evaporation) over each of 190 grid squares, each 40 km × 40 km covering the whole of Britain.
(iii) Calculation of actual evaporation using a soil-moisture extraction model which takes into account the soil type, the vegetation and its state of growth, the surface albedo and the aerodynamic and surface resistances to heat and water vapour transfer.
(iv) Calculation of water balance and excess rainfall.
(v) Data output, which is available from the Meteorological Office in weekly maps and tabular form, of potential and actual evaporation, soil-moisture deficit and *hydrologically effective rainfall* (*HER*), for 3 soils and 13 crops/surface types, together with the meteorological variable. *HER* is the excess rain available after the soil is at field capacity, i.e. the water contributing to stream flow and groundwater.

The system is run on the Meteorological Office computers and regular users of the information can access it by mail and in a limited form, by telex, fax and Prestel.

Figure 4.8 shows the average effective *s.m.d.* (mm) over the British Isles.

Reference to figure 4.8 will show that the likelihood of a flood condition (for example) arising varies greatly from region to region. The average values shown

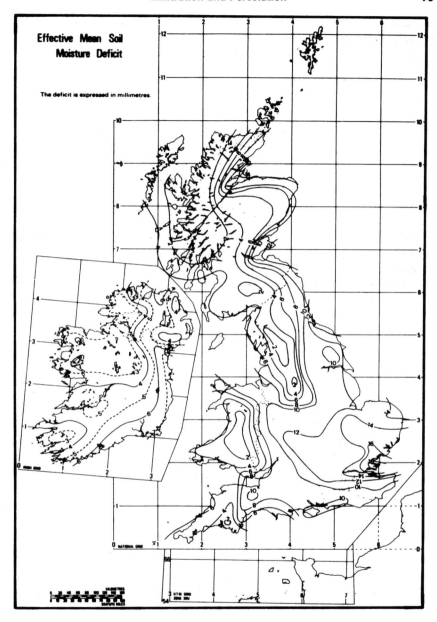

Figure 4.8 *Effective mean soil-moisture deficit in mm for Britain and Ireland (Meteorological Office: Crown copyright)*

are used in chapter 7 for the conversion of gross rainfall to rainfall excess in the estimation of floods on ungauged catchments.

The use of estimated s.m.d. to predict the proportion of runoff arising from particular storms is similar to that for antecedent precipitation index in that the deficit (or index) is deducted from actual precipitation, to give net rain.

4.4.4 Catchment wetness index (CWI). The property of catchment wetness has an important influence on the quantity of net or excess rainfall that provides runoff and in subsequent chapters is used in the estimation of particular flood events by the FSR methods. It is convenient, therefore, to define the index here.

A period of 5 days preceding a storm event is held to provide a recent history of rainfall and a decay index of 0.5 is proposed in the formula for API:

$$\text{API5} = 0.5^{\frac{1}{2}} \left[P_{d-1} + 0.5\, P_{d-2} + (0.5)^2\, P_{d-3} + (0.5)^3\, P_{d-4} + (0.5)^4\, P_{d-5} \right]$$

Here P denotes daily rainfall and the suffix indicates the relevant day. From this

$$\text{CWI} = 125 + \text{API5} - \text{SMD}$$

For the purposes of this text, it is sufficient to know how to derive average values of CWI for use in flood-prediction calculations where the 5 days preceding rainfall is unknown. Such average values have been calculated in the FSR and are strongly dependent on annual average rainfall. The relationship for the British Isles is shown in figure 4.9.

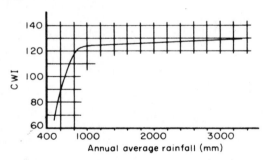

Figure 4.9 *Recommended design values for catchment wetness index
(source: Flood Studies Report)*

4.4.5 Soil classification. A further and perhaps most important property of the catchment in respect of runoff is the soil cover, its depth, permeability and slope.

The Soil Surveys of England and Wales, Scotland, Ireland and the Northern Ireland Ministry of Agriculture have prepared soil maps of the British Isles, which have been used as a basis for calculating a soil index as a term in many of the equations recommended in the Flood Studies Report.

The basis for comparison is 'the winter rain acceptance potential', which is the reverse of the runoff potential. It is influenced by permeability, position of the groundwater level and the slope of the terrain. Five classes of winter rain acceptance are indicated on the soil maps:

Class	Winter rain acceptance	Runoff
1	Very high	Very low
2	High	Low
3	Moderate	Moderate
4	Low	High
5	Very low	Very high

A soil index is calculated from the proportions of a catchment's area that are covered by each class. The precise definition of this index, *SOIL*, is given in section 9.4.1.

The guidelines for each classification are as set out in table 4.2, which can be used for preliminary estimates of ungauged catchments when soil maps are not available.

Soil classification maps (entitled RP, denoting runoff potential) for the British Isles are given in appendix A.

4.4.6 Measurement of soil moisture. An instrument for *in situ* measurement of soil moisture, the Wallingford soil-moisture probe, has been developed by the United Kingdom Institute of Hydrology in conjunction with the Atomic Energy Authority. The instrument is designed for use in the field in all kinds of terrain and in any weather.

The instrument consists of a radioactive source in a 740 mm long probe that can be lowered into an aluminium access tube permanently installed in the ground, a shield and housing for the probe, a suspension cable and meter. Fast neutrons emitted from the radioactive source are scattered and slowed by collisions with the atomic nuclei of soil constituents, mainly by the hydrogen of water in the soil, thus producing 'slow' neutrons. These are sensed by a slow-neutron detector in the probe and converted into electrical pulses, which pass through the suspension cable to a meter where a visual display indicates the rate of detection. The wetter the soil the larger the number of collisions and hence of slow neutrons detected. The displayed detection rate is therefore a function of the soil moisture at the probe depth. The moisture value indicated represents the mean value for a somewhat indefinite 'sphere of influence' within the soil surrounding the source, with a radius that may be regarded as being about 150–300 mm. A series of readings is taken down the profile at intervals usually of 100 mm or 150 mm. The weighted mean of these readings gives the total moisture storage in a given section of the profile.

TABLE 4.2 Classification of soils by runoff potential (source: Flood Studies Report)[a]

Drainage class	Depth to impermeable layer (cm)	Permeability of overlying soil								
		rapid, slope 0–2°	medium, slope 0–2°	slow, slope 0–2°	rapid, slope 2–8°	medium, slope 2–8°	slow, slope 2–8°	rapid, slope more than 8°	medium, slope more than 8°	slow, slope more than 8°
Rarely water-logged within 60 cm (well and moderately well drained)	>80	1	1	2	1	2	2	1	2	3
	40–80	1	1	2	2	2	3	3	3	4
	<40	NOT APPLICABLE								
Commonly water-logged within 60 cm during winter (imperfect and poor drainage)	>80	2	2	3	3	3	4	N/A	4	5
	40–80	2	3	3	3	4	4	4	4	5
	<40	3	4	4	4	4	4	4	5	5
Commonly water-logged within 60 cm winter and summer (very poorly drained)	>80	4	4	5	5	5	5	N/A	5	5
	40–80	4	5	5	5	5	5	N/A	5	5
	<40	5	5	5	5	5	5	5	5	5

[a] Upland and peaty soils are Class 5; urban areas unclassified. Class 1, very low runoff; Class 5, very high runoff.

If a catchment is provided with a number of permanently installed aluminium tubes at suitable locations, then a single battery-powered instrument can be carried from site to site to measure the *in situ* soil moisture at each one, thus removing much of the guess-work from the evaluation of this parameter.

The neutron probe is normally used for making repeated measurements of moisture *change* at the same site and depth and for such a purpose can give a very high precision. The accuracy of absolute moisture values on the other hand is heavily dependent on accurate calibration at each site and depth and this is generally impracticable.

A diagram of the instrument and a photograph of one in use are shown in figures 4.10 and 4.11 respectively.

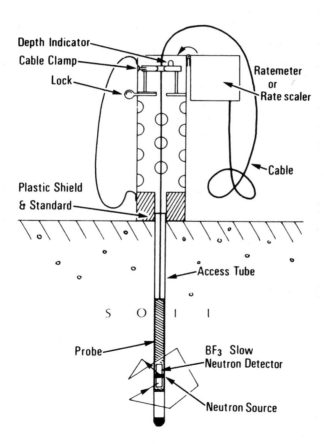

Figure 4.10 *Wallingford soil-moisture probe (courtesy of the Institute of Hydrology)*

Figure 4.11   *Wallingford soil-moisture probe*

References

1. NASSIF, S. H. and WILSON, E. M. The influence of slope and rain intensity
    on runoff and infiltration. *Bull. Int. Assoc. Sci. Hydrol.*, **20**, No. 4 (1976).
2. HORTON, R. E. The role of infiltration in the hydrologic cycle. *Trans. Am.
    Geophys. Union*, **14**, ((1933) 443–60

3. BOUCHARDEAU, A. and RODIER, J. Nouvelle Méthode de détermination de la capacité d'absorption en terrains perméables. *La Houille Blanche*, No. A (July/August 1960) 531–6
4. SOR, K. and BERTRAND, A. R. Effects of rainfall energy on the permeability of soils. *Proc. Am. Soc. Soil Sci.*, **26**, No. 3 (1962)
5. HORTON, R. E. Determination of infiltration capacity for large drainage basins. *Trans. Am. Geophys. Union*, **18**, (1937) 371
6. SHERMAN, L. K. Comparison of F-curves derived by the methods of Sharp and Holtan and of Sherman and Mayer. *Trans. Am. Geophys. Union*, **24**, No. 2 (1943) 465
7. BUTLER, S. S. *Engineering Hydrology*, Prentice-Hall, Englewood Cliffs, New Jersey, 1957
8. LINSLEY, R. K., KOHLER, M. A. and PAULHUS, J. L. H. *Hydrology for Engineers*, McGraw-Hill, New York, 1958, p. 162
9. Estimated soil moisture deficit over Gt. Britain. *Explanatory Notes Meteorological Office, Bracknell* (issued twice monthly)

Further reading

BELL, J. P. Neutron probe practice. *Report No. 19, Institute of Hydrology*, Wallingford, United Kingdom
GRINDLEY, J. Estimation of soil moisture deficits. *Meteorological Magazine*, **96**, (1967) 97
GRINDLEY, J. Estimation and mapping of evaporation. *Int. Assoc. Sci. Hydrol. Symposium on World Water Balance, Reading. IASH Publication 92*, 1970, pp. 200–213
HORTON, R. E. Analyses of runoff-plot experiments with varying infiltration capacity. *Trans. Am. Geophys. Union*, Part IV, (1939) 693
PENMAN, H. L. The dependence of transpiration on weather and soil conditions, *J. Soil Sci.*, **1**, (1949) 74
WILM, H. G. Methods for the measurements of infiltration. *Trans. Am. Geophys. Union*, Part III, (1941) 678

Problems

4.1 Discuss the influence of slope of catchment and rainfall intensity on infiltration rates under constant rainfall.

4.2 Discuss the influence of forestation and agriculture on groundwater. Present arguments for and against

(a) livestock rearing
(b) crop growing
(c) forestation

on the catchment of a public water-supply reservoir.

4.3 The table below gives the hourly rainfall of three storms that gave rise to runoff equivalent of 14, 23 and 18.5 mm respectively

Hour	Storm 1 (mm)	Storm 2 (mm)	Storm 3 (mm)
1	2	4	3
2	6	9	8
3	7	15	11
4	10	12	4
5	5	5	12
6	4		3
7	4		
8	2		

Determine the Φ-index for the catchment.

4.4 Why is the method of subtracting infiltration rates from rainfall intensities to compute hydrographs of runoff not applicable to large natural river basins?

4.5 The antecedent precipitation index for a station was 53 mm on 1 October; 55 mm rain fell on 5 October, 30 mm on 7 October and 25 mm on 8 October. Compute the antecedent precipitation index I_t

(a) for 12 October, if $k = 0.85$
(b) for same date assuming no rain fell.

4.6 Use the co-axial relationship of figure 4.7 to determine how the runoff in this river changes seasonally. Assume that during week number 1 a storm of 5 in. of rain lasting 72 h occurs. Compare what happens with the effects of the same storm in week number 25, if the antecedent precipitation index in each case is 1.5 in. Suggest which seasons of the year the weeks are in and explain why there should be a difference in runoff.

4.7 (a) Describe 'antecedent precipitation index' and 'soil moisture deficit', indicating how they are derived and used.
(b) List the parameters necessary for the solution of Penman's equation for open-water evaporation. Describe how you would obtain suitable values for each of them.

4.8 Calculate the Catchment Wetness Index (CWI) for a point 400 km north and 400 km east of the UK National Grid origin, if the preceding 5 days' rainfall were (in mm)

$$12.0, 0.5, 4.2, 0.0, 3.5$$

Compare this with the FSR design values.

5　Groundwater

5.1 The occurrence of groundwater

Rainfall that infiltrates the soil and penetrates to the underlying strata is called *groundwater*. The quantity of water that can be accommodated under the surface depends on the porosity of the sub-surface strata. The water-bearing strata, called *aquifers*, can consist of unconsolidated materials like sands, gravels and glacial drift or consolidated material like sandstones and limestones. Limestone is relatively impervious but is soluble in water and so frequently has wide joints and solution passages that make the rock, *en masse* similar to a porous rock in its capacity to hold water and act as an aquifer.

The water in the pores of an aquifer is subject to gravitational forces and so tends to flow downwards through the pores of the material. The resistance to this underground flow varies widely and the *permeability* of the material is a measure of this resistance. Aquifers with large pores such as coarse gravels are said to have a high permeability, and those with very small pores such as clay, where the pores are microscopic, have a low permeability.

As the groundwater percolates down, the aquifer becomes saturated. The surface of saturation is referred to as the groundwater table or the *phreatic surface*. This surface may slope steeply and its stability is dependent on supply from above. It falls during dry spells and rises in rainy weather. The water in the aquifers is usually moving slowly towards the nearest free water surface such as a lake or river, or the sea. However, if there is an impermeable layer underlying an aquifer and this layer outcrops on the surface, then the groundwater will appear on the surface in a seepage zone or as a *spring*. It is equally possible for a groundwater aquifer to become overlain by impermeable material and so be under pressure. Such an aquifer, fed from a distance, is called a *confined aquifer* and the surface to which the water would rise if it could is called the *piezometric surface*. Another name, used for wells drilled into such confined aquifers, is *artesian wells*, and the word *artesian* is sometimes applied also to the aquifers. If the piezometric surface is above ground level at an artesian well, it is called a flowing well, and a fracture or flaw in the impermeable overlay will, in such conditions, result in an *artesian spring*. Sometimes a small area of impermeable material exists in a large aquifer. This happens through geological faulting or, for example, through a lens of clay occurring in otherwise sandy glacial drift. A small local water-table, called a *perched water-table* may result and this can often be a long way above the true phreatic surface.

Some of the modes of occurrence of groundwater described above are illustrated in figure 5.1.

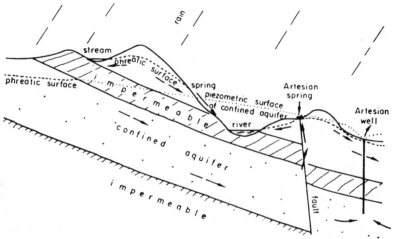

Figure 5.1   *Modes of occurrence of groundwater*

5.2 Factors of influence

The flow of groundwater takes place in porous media. The pores through which movement takes place can be very small indeed and generally are between the limits of 2 and 0.02 mm. The movement is slow by the standard of surface runoff and the flow is usually *laminar*. The Reynolds number in flow of this kind is very low.

The factors of importance in the flow are

(1) the liquid—its density and viscosity,
(2) the media through which the liquid moves,
(3) the boundary conditions.

The liquid normally is water, usually fresh but occasionally saline. Its temperature may vary within the range 0–30 °C.

Density.   The density of fresh water varies only very slightly with temperature and its effect can be ignored in groundwater flow:

Temperature ($^\circ$C)	0	4	10	15	20
Density ρ (g/l)	999.868	1000	999.727	999.126	998.230

Of greater importance is salinity, the effect of which is shown in figure 5.2.

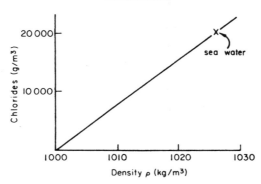

Figure 5.2 *Density of saline water*

Viscosity is a measure of the shear strength of a liquid; the lower the viscosity, the more mobile or penetrating the liquid.

Absolute viscosity, denoted by μ, has units of N s/m^2; the CGS unit, the poise = 10^{-1} N s/m^2. Water at 20°C has a viscosity of 1 *centipoise* (0.01 poise = 10^{-3} N s/m^2).

Kinematic viscosity, denoted by ν, is the ratio of absolute viscosity to density, or $\nu = \mu/\rho$, and has units of m^2/s; the CGS unit, the *stoke* = 10^{-4} m^2/s. Kinematic viscosity occurs in many applications; for example, Reynolds number, $R = (\nu D)/\nu$.

$$\nu(\text{water}) \approx 10^{-1}\ \text{m}^2/\text{day} \approx 10^{-6}\ \text{m}^2/\text{s} \approx 10^{-2}\ \text{stoke}$$

The kinematic viscosity of liquids is almost independent of pressure and is substantially a function of temperature. For water

Temperature (°C)	0	5	10	15	20
ν(m^2/day)	0.152	0.133	0.113	0.098	0.087

The media in which groundwater moves are characterised by the properties of porosity, permeability and, to a minor extent, compressibility. Only the first two are considered here.

Porosity is defined as n = total voids/total volume and ranges from a few percent to about 90 per cent. In a granular mass composed of perfect uniform spheres:

in the loosest possible packing, n = 47.6 per cent
in the densest possible packing, n = 26 per cent

Natural soils are, of course, composed of irregular particles of many different sizes. The more regular the soil, the more porous it tends to be, since in irregular soils the smaller particles tend to fill the voids in the larger particle packing. It is therefore, standard procedure in any groundwater survey to analyse the soils mechanically and plot the particle sizing in a standard way, using a logarithmic size scale. A typical analysis is shown in figure 5.3. Two soils are plotted: the more regular soil has the steeper slope and is likely to be more porous.

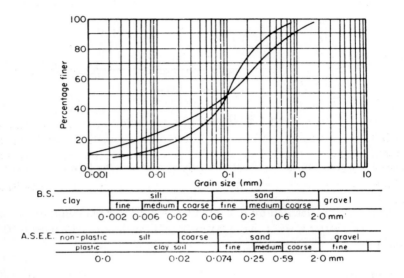

Figure 5.3 *Plotting of particle size analysis, and nomenclature*

When water fills the pores of a soil there is a thin layer, only a few molecules thick, that coats the particles. This water is not free to move and adheres to the particles even when the voids have been drained, occupying part of the available space. This means that the *effective porosity*, n_e, may be less than the true porosity n. In coarse materials such as gravels there will be no difference between n_e and n but in fine sands the difference may be 5 per cent, or even more in very fine materials. In most considerations of porosity in the flow of groundwater it is n_e, the effective porosity, that is of importance.

Permeability is a function of porosity, structure, and the geological history of the material. By structure is meant the grain size, distribution, orientation, arrangement and shape of the particles. For example, in a sediment of predominantly flat grains deposited in water, the grains will tend to lie with their longest axes horizontal. In such a soil the permeability may well be higher in a horizontal direction than in a vertical one. Such a soil is called *anisotropic*.

The permeability of a particular material is defined by its *permeability coefficient*, denoted by k; k depends on the factors listed above, which may be described as the geometry of the pore system, and is expressed in metres per day, or feet per day.

Many attempts have been made to find a formula for k in terms of measurable values of the properties of the material. Generally this is very difficult and such formulae can be used with confidence only in strictly limited applications. For example, a formula used in connection with sand filters for water supply, and applying to only homogeneous rounded grain media, of not too fine a size, is

$$k = Cd_{10}^2$$

where k = permeability coefficient in m/day

d_{10} = the grain size in mm where 10 per cent of material is finer and 90 per cent coarser

C = a constant of value 400–1200 (an average value is 1000)

It must be emphasised that formulae such as this are of little value in materials of a heterogeneous character or outside their precisely defined limits; k is not necessarily a constant for a particular soil, as chemical erosion or deposition can sometimes occur with percolating groundwater.

Some values of k as they occur naturally are indicated on the logarithmic scale of permeability in figure 5.4.

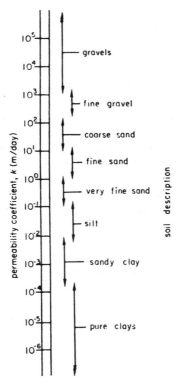

Figure 5.4 *Range of permeability in natural soils*

5.3 Groundwater flow

5.3.1 Darcy's Law. Before any mathematical treatment of the flow of ground-water can be attempted, it is necessary to make certain simplifying assumptions

 (1) the material is homogeneous and isotropic,
 (2) there is no capillary zone,
 (3) there is a steady state of flow.

The fundamental law is Darcy's Law of 1856. This states that the rate of flow per unit area of an aquifer is proportional to the gradient of the potential head measured in the direction of flow. Or

$$v \propto i$$

By introducing a constant of proportionality, which is k, the permeability coefficient

$$v = ki$$

and for a particular aquifer or part of an aquifer of area A (area at right angles to flow)

$$Q = vA = kAi$$

where v = velocity of the water (measured as the time taken to pass between two reference points) in m/day (or m/s or ft/s etc.) and called the *specific velocity*

i = the hydraulic gradient—this equals the potential gradient since velocities are so small that there is virtually no velocity head; i is written also as $d\phi/dl$, l being distance along a flow line, and ϕ the potential head.

The specific velocity is not the true velocity, but is merely Q/A. The actual velocity of water in the pores is greater than the specific velocity since the path the water follows through porous media is always longer than a straight line beween any two points.

If the average real, or pore, velocity is denoted by $\bar{v}$ then

$$\bar{v} = \frac{\text{discharge}}{\text{area of water passage}} = \frac{Q}{An_e} = \frac{Av}{An_e} = \frac{v}{n_e}$$

Hence

pore velocity (average) = specific velocity ÷ effective porosity

Since the velocity distribution across a pore is probably parabolic, being highest in the centre and zero at the edges, the maximum pore velocity v_{max} = $2 \times \bar{v}$ (approximately).

So in a typical case where (say) $n_e = \frac{1}{3}$, then

$$\bar{v} = 3v \quad \text{and} \quad \bar{v}_{max} = 6v$$

While these are typical figures only, it is important to remember the order of the velocities, since it is on $\bar{v}_{max}$ that the Reynolds number and the continuance of laminar flow depends.

5.3.2 Flow in a confined aquifer. Consider now the case of unidirectional flow in a confined aquifer of permeability k, illustrated in figure 5.5. Groundwater is

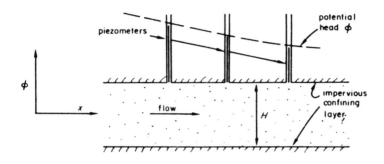

Figure 5.5 *Flow in a confined aquifer*

flowing from left to right, the energy required to move the water through the pores is continually using up the available head and so the line of potential head as indicated by piezometers introduced into the aquifer, is declining. From Darcy's Law

$$v_x = -k \frac{d\phi}{dx}$$

and if q = flow in the aquifer per unit width then

$$q = -kH \frac{d\phi}{dx} \tag{5.1}$$

Since it is assumed that the flow is in a steady state

$$\frac{dq}{dx} = 0$$

and differentiating the equation for q, above

$$\frac{dq}{dx} = -kH \frac{d^2\phi}{dx^2}$$

So

$$\frac{d^2\phi}{dx^2} = 0 \qquad (5.2)$$

since both k and H have finite values.

Equations 5.1 and 5.2 are the fundamental differential equations for this case of a confined aquifer. By introducing suitable boundary conditions, many problems in this case can be solved by using these equations.

Note that the v_x in the Darcy equation is the specific velocity.

5.3.3 Flow in an aquifer with phreatic surface. Consider now the case of an aquifer with a phreatic surface, resting on an impermeable base. This case is illustrated in figure 5.6. Here the first equation from Darcy's Law would be

$$v_s = -k \frac{d\phi}{dl}$$

where l = distance measured in the direction of flow.

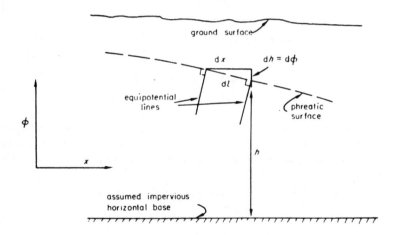

Figure 5.6 *Flow in an aquifer with phreatic surface*

If two assumptions (due to Dupuit) are made

1. that $d\phi/dl$ can be represented by $d\phi/dx$ (permissible if $d\phi$ is small), and
2. that all flow lines in the aquifer are horizontal and equipotential lines vertical (nearly true except near abstraction points) so that $d\phi/dx = dh/dx$ then the Darcy equation becomes

$$q = -kH \frac{dh}{dx} \qquad (5.3)$$

and

$$\frac{dq}{dx} = -\tfrac{1}{2}k \frac{d^2(h^2)^*}{dx^2}$$

and since by continuity $dq/dx = 0$, then

$$\frac{d^2(h^2)}{dx^2} = 0 \tag{5.4}$$

Equations 5.3 and 5.4 are the fundamental equations for solving problems in the case of phreatic aquifers (except where Dupuit's assumptions are no longer reasonable).

If the aquifer is being recharged by rain falling on the ground surface, let the net infiltration rate be N in suitable units (as figure 5.7).

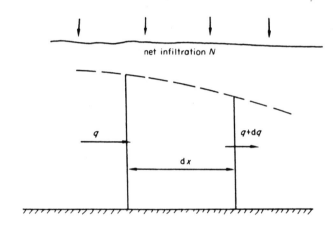

Figure 5.7 *Flow in a phreatic aquifer with rainfall*

---

* The power rule of differentiation may be written $d(v^n)/dx = nv^{n-1}. dv/dx$. Hence

$d(h^2)/dx = 2h. dh/dx$

and therefore

$\tfrac{1}{2} d(h^2)/dx = h. dh/dx$

So since

$q = -kh. dh/dx$

therefore

$q = -\tfrac{1}{2}k. d(h^2)/dx$

and hence

$dq/dx = -\tfrac{1}{2}k. d^2(h^2)/dx^2$

In this case $dq = N.dx$, so therefore

$$\frac{dq}{dx} = -\tfrac{1}{2}k \frac{d^2(h^2)}{dx^2} = N$$

and hence

$$\frac{d^2(h^2)}{dx^2} = -\frac{2N}{k} \tag{5.5}$$

Now equations 5.3 and 5.5 are relevant.

Example 5.1. Suppose there are two canals, at different levels, separated by a strip of land 1000 m wide, of permeability $k = 12$ m/day as shown in figure 5.8. If one canal is 2 m higher than the other and the depth of the aquifer is 20 m below the lower canal to an impermeable base, find the inflow into, or abstraction from, each canal per metre length. Take annual rainfall as 1.20 m per annum and assume 60 per cent infiltration.

Assume a reference origin as indicated on figure 5.8. Then the boundary conditions are simply: when $x = 0$, $h = 20$; and when $x = 1000$, $h = 22$.

$$N = 60 \text{ per cent of } 1.2 = 0.72 \text{ m/year}$$
$$= 0.72/365 \text{ m/day}$$

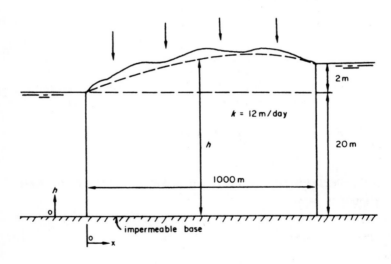

Figure 5.8   *Flow between two canals*

From equation 5.5

$$\frac{d^2(h^2)}{dx^2} = -\frac{2N}{k}$$

Integrating twice

$$\frac{d(h^2)}{dx} = -\frac{2N}{k}x + C_1$$

$$h^2 = -\frac{N}{k}x^2 + C_1 x + C_2 \tag{5.6}$$

When $x = 0, h = 20$, therefore

$$400 = 0 + 0 + C_2$$

and

$$C_2 = 400$$

When $x = 1000, h = 22$, therefore

$$484 = -\frac{0.72 \times 10^6}{365 \times 12} + 1000 C_1 + 400$$

and

$$C_1 = 0.084 + 0.164 = 0.248$$

Now from equation 5.3

$$q = -kh\frac{dh}{dx}$$

and from equation 5.6

$$h = \sqrt{\left(-\frac{N}{k}x^2 + 0.248x + 400\right)}$$

Let the expression under the square root sign $= u$, then $h = u^{\frac{1}{2}}$, and therefore

$$\frac{dh}{dx} = \frac{1}{2u^{\frac{1}{2}}}\left(-\frac{N}{k}.2x + 0.248\right)$$

At $x = 0$

$$q = -ku^{\frac{1}{2}}.\frac{1}{2u^{\frac{1}{2}}}(0.248) = -6(0.248) = -1.49 \text{ m}^3/\text{day/m}$$

At $x = 1000$

$$q = -\frac{k}{2}\left(-\frac{2000 \times 0.72}{365 \times 12} + 0.248\right)$$

$$= -6(-0.328 + 0.248) = 6(0.08) = 0.48 \text{ m}^3/\text{day/m}$$

So there is discharge into both canals from the aquifer of 1.49 m^3/day to the lower, and 0.48 m^3/day to the upper, for each metre length of aquifer.

The foregoing simple cases will serve to show the way in which the movement of groundwater can be analysed. As conditions become more complex, the solutions become more difficult, but standard solutions have been computed for very many groundwater situations and most conditions met with in nature can be analysed, at least approximately.

5.4 The abstraction of groundwater

The simplest and oldest way of collecting groundwater is by digging a hole in the ground that penetrates the water-table. If the quantity of water that can be taken from the hole is not adequate, then the hole must be extended either horizontally or vertically. Which method is chosen will depend on the local geohydrology.

If the hole is extended horizontally it becomes an open collecting ditch. Alternatively, it could be underground as a collecting tunnel. These horizontal collectors must be used if the aquifer thickness is small and if the drawdown due to abstraction has to be limited; for example, when a layer of fresh water overlies a layer of salt water.

The vertical extension of the hole makes it a dug or drilled well, or a borehole. This method can be used when the aquifer is of sufficient thickness, and in any case when the aquifer is more than about 6 m below ground level. Dug wells are usually 1 m or more in diameter and so the shaft acts as a reservoir for a short-term, high-rate abstraction. The large diameter is also useful when the entrance velocity of water into the shaft has to be kept low; for example, in fine-grained sands.

The majority of wells sunk nowadays for water supply are drilled wells and these are commonly from 30 m to 500 m deep. They are constructed by using drill bits that break the material at the bottom of the hole into small pieces, so that these can be removed with other tools. Two principal methods are employed: percussion drilling and rotary drilling. The percussion method involves alternately raising and dropping the tools in the borehole; in the rotary method, a rotating bit cuts or abrades the hole bottom. Drilled wells can penetrate any materials from soft clay to hard rock up to depths of a kilometre or more.

As the well is drilled it is 'cased' with steel piping to prevent wall collapse. At the bottom of the casing a well screen is constructed. This is the point where groundwater enters and the screen is necessary to prevent the washing in of fine particles and consequent clogging of the well bottom and pump. The screen should cause as small a loss of head as possible and be structurally strong, corrosion resistant and reasonably cheap. These requirements are to some extent contradictory, since the smaller the screen openings and thus the more effective they are in keeping out fine particles, the greater the head loss that will be caused.

Modern well screens are usually slotted with fine slots in a plastic material, though steel, copper, bronze, wood, vitrified clay and glass are used. Gravel packs are customarily placed around the screen to act as preliminary filters; in some cases gravel packs that consist of particles of decreasing diameter, placed in concentric rings, can be used with a simply perforated bottom pipe on the casing. Figure 5.9 shows a modern PVC slotted-tube well-screen, available in diameters between 56 mm and 200 mm in standard lengths of 3 m.

Figure 5.9 *A modern PVC well-screen (courtesy of Boode B.V., Zevenhuizen, The Netherlands)*

The construction of drilled wells, well screens and gravel packs and the techniques of well development and maintenance are beyond the scope of this book. Readers requiring information on these topics should consult references [1] and [2].

Once the water has entered the well it has to be pumped to the surface. Well pumps are classified as reciprocating, rotating vertical shaft, jet and air lift pumps. Rotating vertical shaft are either surface driven or submersible, and can be centrifugal or rotary positive displacement.

By far the most common application is now the electric submersible centrifugal borehole pump, with the electric drive motor directly coupled to the pump stages in one long pump body that is placed near the bottom of the well. Such pumps are manufactured in sizes down to 100 mm diameter to supply heads up to 100 m or more if necessary. Such a 100 m (4 in.) diameter pump would supply about 4 m^3/h while a 250 mm (10 in.) diameter pump might supply 30 times as much. A sketch of a typical installation is shown in figure 5.10.

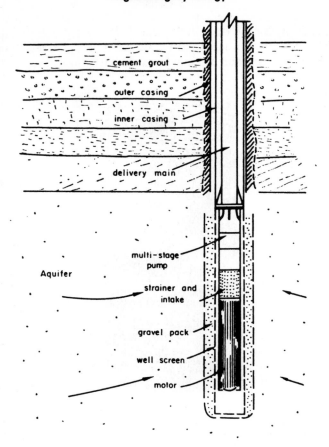

Figure 5.10 *Sketch of bottom of a typical water-supply borehole*

5.5 The yield of wells

Formulae for the drawdown curves of a single well can be derived from the conditions of flow previously discussed in section 5.3. Only the two simplest cases will be considered here: (1) steady flow to a well tapping confined ground-water, and (2) steady flow to a water-table well.

Many other factors should be considered (for example, the influence of partial penetration of the well in the aquifer, unsteady flow etc.) but for a full treatment of these the reader should consult references 1 and 2 at the end of this chapter.

5.5.1 Steady confined flow. Drawdown is denoted by s and is measured from the undisturbed piezometric surface before pumping (see figure 5.11). The horizontal co-ordinate is measured radially from the well and is denoted by r, since the flow is radial to the well. The steady-state discharge from the well is called Q_0.

The equations governing flow are written as follows: from Darcy's Law

$$Q = vA = -k \frac{ds}{dr}.2\pi rH$$

and from continuity

$$Q = Q_0 = \text{constant}$$

Combining these

$$ds = -\frac{Q_0}{2\pi kH}.\frac{dr}{r}$$

Integrating between the limits $r = r_1, s = s_1$, and $r = r_2, s = s_2$

$$s_1 - s_2 = \frac{Q_0}{2\pi kH}.\ln\frac{r_2}{r_1} \text{ (Theim's equation)} \qquad (5.7)$$

This is a very important equation and holds true (at least approximately) for all kinds of underground flow, steady and unsteady, confined and unconfined. The value of kH is known as the *coefficient of transmissibility*.

Equation 5.7, indefinitely integrated, yields

$$s = -\frac{Q_0}{2\pi kH}.\ln r + C$$

and if $s = 0$ when $r = R$ then

$$s = \frac{Q_0}{2\pi kH}\ln\frac{R}{r} \qquad (5.8)$$

Either of the two equations 5.7 and 5.8 enables the drawdown curve to be established, provided that the integration constants Q_0 and R can be determined from the boundary conditions. Q_0 is the constant discharge of the pumped well and so can be measured. R, however, varies from one observation point to another. In most cases, however, it is the drawdown close to the well that is important, where R has the value R_0. Some values of R_0 for idealised boundaries are given in figure 5.12, where the shaded areas are land and the blank adjacent spaces are open water.

This figure illustrates how R_0 depends on the distance to open water. As the distance increases, so does R_0 and the drawdown. Indeed finite drawdowns are possible only when the constant-level open water is a finite distance away. Also,

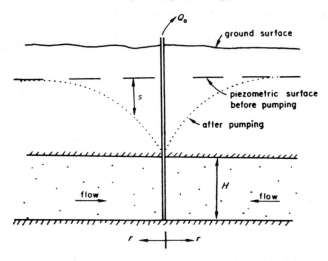

Figure 5.11 *Well pumping from a confined aquifer*

there is no great change in R_0 even for widely different boundary conditions, and since it is the natural logarithm of the ratio R_0/r that affects drawdown, an informed estimate of R_0 will often give adequately accurate results.

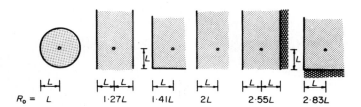

Figure 5.12 *Value of the integration constant R_0 in Dupuit's formula, for drawdown close to the well* [3]

Example 5.2. A fully-penetrating well, with an outside diameter of 0.5 m, discharges a constant 0.08 m³/s from an aquifer whose coefficient of transmissibility is 25×10^{-3} m²/s. The aquifer is in contact with a lake 2 km away and has no other source of supply. Estimate the drawdown at the wellface.

$$s = \frac{Q_0}{2\pi kH} \cdot \ln \frac{R_0}{r}$$

Since R_0 may be taken as $2L$ (= 4000 m) from figure 5.12, then

$$s_0 = \left(\frac{0.08}{2\pi .25 \times 10^{-3}} \right) \ln \frac{4000}{0.25} = \frac{0.08 \times 9.68 \times 10^3}{157.08}$$

Hence the drawdown at the wellface is 4.9 m.

5.5.2 Steady unconfined flow. When the drawdown is slight compared with the thickness of the aquifer, the factor kH remains nearly constant and the formula for steady confined flow may be used. As drawdown increases, the falling water level reduces the area of transmitting aquifer and the equations now become, in the notation of figure 5.13: from Darcy's Law

$$Q = 2\pi r . h . k \frac{dh}{dr}$$

and from continuity

$$Q = Q_0 = \text{constant}$$

Combining these

$$h . dh = \frac{Q_0}{2\pi k} . \frac{dr}{r}$$

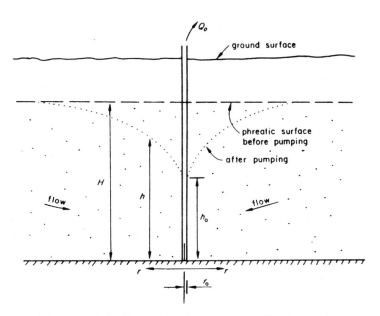

Figure 5.13 *Well pumping from an unconfined aquifer*

Integrating

$$h^2 = \frac{Q_0}{\pi k} \ln r + C$$

and if

$$h = H \text{ at } r = R$$

then

$$H^2 - h^2 = \frac{Q_0}{\pi k} \ln \frac{R}{r} \tag{5.9}$$

The value of R must satisfy the boundary conditions. Then the drawdown at the wellface $(H - h_0)$ is deduced by introducing the radius of the well r_0 since

$$H^2 - h_0^2 = \frac{Q_0}{\pi k} \ln \frac{R}{r_0}$$

Example 5.3. A well is drilled to the impermeable base in the centre of a circular island of 1 mile diameter in a large lake. The well completely penetrates a sandstone aquifer 50 ft. thick overlain by impermeable clay. The sandstone has a permeability of 50 ft./day. What will be the steady discharge if the drawdown of the piezometric surface is not to exceed 10 ft. at the well, which has a diameter of 1 ft?

For a well in the centre of the circular island, the boundary condition is $s = 0$ when $r = 2640$ ft.

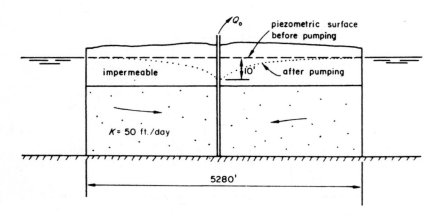

Figure 5.14 *Pumping from a central well in a circular island and confined aquifer*

Then from equation 5.8

$$10 = \frac{Q_0}{2\pi \times 50 \times 50} \ln \frac{2640}{0.5}$$

Therefore

$$Q_0 = \frac{50000\pi}{\ln 5280} = \frac{157080}{8.572} = 18340 \text{ ft}^3/\text{day}$$

$$= 0.212 \text{ ft.}^3/\text{s}$$

5.5.3 Steady unconfined flow with rainfall. When rainfall is present the equations become: from Darcy's Law

$$Q = 2\pi r h \cdot k \frac{dh}{dr}$$

and from continuity

$$dQ = -2\pi r \cdot dr \cdot N \text{ (where } N = \text{net infiltration)}$$

Integrating

$$Q = -\pi r^2 N + C_1$$

and C_1 may be determined from the condition that when $r = r_0 \approx 0$, $Q = Q_0$. Therefore

$$Q = -\pi r^2 N + Q_0$$

Substituting this value in the Darcy equation

$$h \cdot dh = \frac{Q_0}{2\pi k} \cdot \frac{dr}{r} - \frac{N}{2k} \cdot r \, dr$$

and integrating

$$h^2 = \frac{Q_0}{\pi k} \ln r - \frac{N}{2k} r^2 + C_2 \tag{5.10}$$

C_2 is an integration constant that must satisfy the particular boundary conditions. In the case of a central well in a circular island of radius R, when $r = R$, $h = H$, so

$$C_2 = H^2 - \frac{Q_0}{\pi k} \ln R + \frac{N}{2k} R^2$$

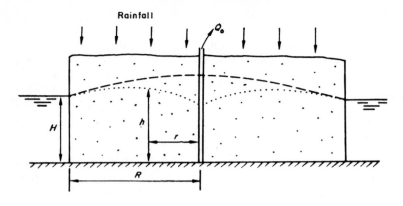

Figure 5.15 *Central well in a circular island and unconfined aquifer with rainfall*

Substituting this value in equation 5.10

$$H^2 - h^2 = \frac{Q_0}{\pi k} \ln \frac{R}{r} - \frac{N}{2k} (R^2 - r^2)$$

If $Q_0 = 0$ (that is, there is no pumping) then the shape of the phreatic surface is given by

$$H^2 - h^2 = - \frac{N}{2k} (R^2 - r^2) \qquad (5.11)$$

Example 5.4. A circular island 500 m radius has an effective infiltration N of 4 mm/day. A central well is pumped to deliver a constant Q_0 of 25 m³/h from an aquifer with dimensions and properties as shown in figure 5.16. What is the drawdown at the well and at the water divide?

(1) Assume no pumping. Then from equation 5.11

$$H^2 - h_1^2 = - \frac{N}{2k} (R^2 - r^2)$$

$$100 - h_1^2 = - \frac{0.004}{40} (250000 - r^2)$$

$$100 - h_1^2 = - 25 + \left(\frac{r}{100}\right)^2 \qquad (5.12)$$

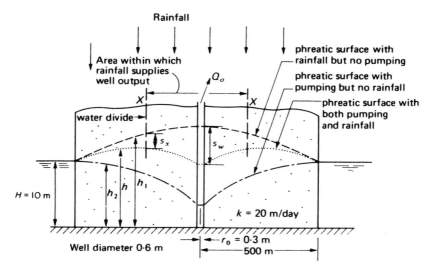

Figure 5.16 *Circular island with central well, rainfall and an unconfined aquifer. Solution by superposition*

(2) Assume no rainfall. Then from equation 5.9

$$H^2 - h_2^2 = \frac{Q_0}{\pi k} \ln \frac{R}{r}$$

$$100 - h_2^2 = 21.9 \log \frac{500}{r} \qquad (5.13)$$

Inserting $r = 0.3$ m (at the wellface) in equations 5.12 and 5.13 yields $h_1 = 11.18$ m and $h_2 = 5.43$ m. Since $h = h_1 + h_2 - H$ (by superposition of drawdowns), it follows that

$$h = 11.18 + 5.43 - 10$$

$$= 6.61 \text{ m}$$

Therefore

$$s_w = h_1 - h$$

$$= 4.57 \text{ m}$$

If a water-divide exists, then all the output of the well is being contributed by rainfall, since if the sea around the island were contributing, the hydraulic gradient would be sloping downwards and inwards at every point. So the area contributing is obtained from

$$Q_0 = \pi r_x^2 N$$

where r_x is radius of the divide, so that

$$25 = \pi r_x^2 \times \frac{0.004}{24}$$

Therefore

$$r_x = 218 \text{ m (well within the 500 m radius of the land)}$$

Using this value of r, equation 5.12 yields $h_1 = 10.97$ and equation 5.13 yields $h_2 = 9.60$. Then

$$h = h_1 + h_2 - H$$
$$= 10.97 + 9.60 - 10.0$$
$$= 10.57$$

So, calling drawdown at the divide s_x

$$s_x = 10.97 - 10.57$$
$$= 0.40 \text{ m}$$

Suppose the simple formula of equation 5.8 had been used, thus assuming constant aquifer thickness. In this case, without rainfall

$$s_w = \frac{Q_0}{2\pi kH} \ln \frac{R}{r} = \frac{600}{2\pi \times 20 \times 10} \times 2.3 \log \frac{500}{0.3}$$
$$= 0.477 \times 2.3 \times 3.223$$
$$= 3.54 \text{ m (compare the value of 4.57 m above)}$$

and

$$s_x = 0.477 \times 2.3 \times \log \frac{500}{218}$$
$$= 1.01 \times 0.36$$
$$= 0.364 \text{ (compare the value of 0.40 m above)}$$

It will be realised that the simple formula for the confined case is adequate in this case except in the immediate neighbourhood of the well. It would, of course, still be necessary to compute the 'no-pumping' phreatic surface, for the case with rainfall.

5.6 Test pumping analysis

5.6.1 General. Using equation 5.8, the drawdown resulting from groundwater abstraction from wells can be determined if the boundary conditions and geo-hydrological constants are known. In particular, the presence of water-bearing strata that form aquifers, their extent, thickness and permeability are usually unknown until a number of test borings have been made. Each test boring, as well as providing information on the underlying geology of an area, may be left open with a simple porous screen in the bottom of the hole to allow subsequent observation of water levels.

Although much information can be gleaned from the drilling of these test bore holes/observation wells, the extent of the yield of an aquifer can finally be determined only by test pumping from a well. Such a well should be positioned so that observation wells are placed on either side of it, on a line through the well, and preferably on two lines at right angles with the pumping well at the intersection.

Pumping tests can be performed either by pumping from the well at a steady rate until steady-state conditions are obtained, (that is, there is no appreciable movement with time in any of the observation wells) and then plotting the data recorded, or by observation of the rate of change of level in all the wells up to and including steady-state conditions. Variants include stepped pumping tests where the discharge is progressively increased at regular intervals [4]. Only the first of these methods is discussed here. For full treatment of groundwater recovery, including test pumping refer to Huisman [3], Verruijt [5], and other specialised texts.

Observation wells should ideally be spaced at increasing intervals from the pumping well—say at 20, 50, 100, 200 and 500 m—depending on the depth and expected productivity of the aquifer. Always use the largest capacity pump available to do the actual pumping. Constant-rate extraction may have to continue for days, or even weeks and months in some cases before steady-state conditions are reached. Careful observation of all the wells should, therefore, be made before starting pumping, at regular intervals throughout the test and during recovery of the levels after pumping has ceased, until the initial equilibrium levels are regained.

5.6.2 Test pumping a well in a confined aquifer. It is assumed there is no supply to the aquifer from above (rainfall) or below. Then, from equation 5.8

$$s = \frac{Q_0}{2\pi kH} \ln \frac{R}{r}$$

where R depends on both boundary conditions and the point of observation. If the aquifer is of large extent and r (the radial distance to an observation well) is not excessive then for $r < 0.1R$

$$s = \frac{Q_0}{2\pi kH} \ln \frac{R_0}{r} \tag{5.14}$$

where R_0 is the integration constant for the well-face, and for all points of observation has the same value.

If the observed values of drawdown s are now plotted against distances r on logarithmic paper, then a straight-line relationship is obtained:

$$s = \left[\frac{Q_0}{2\pi kH} 2.3 \log R_0\right] - \left[\frac{Q_0}{2\pi kH} 2.3\right]\log r$$

or

$$s = A - B \log r$$

The geohydrological constants for the aquifer may now be calculated from

$$kH = \frac{1}{B}\frac{1.15 Q_0}{\pi}$$

and

$$\log R_0 = \frac{A}{B}$$

Example 5.5.   A fully penetrating well in a confined aquifer without recharge is pumped with a constant discharge of 0.03 m³/s until steady conditions obtain. Drawdowns are then obtained from observation wells as indicated below:

s(m)	1.20	1.10	0.81	0.60	0.47	0.31
r(m)	10	20	40	80	120	200

Determine the aquifer formation constants.

From figure 5.16, which is a plot of s versus r

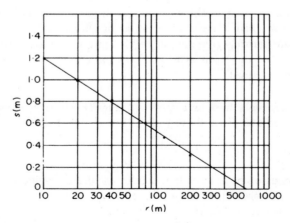

Figure 5.17   *Plot of observed drawdown against distance from pumping well (example 5.5)*

when $\qquad s = 0$: $r = 610$ and $\log r = 2.79$

$\qquad\qquad s = 1.2$: $r = 10$ and $\log r = 1.00$

giving the relationship

$$s = 1.87 - 0.67 \log r$$

From which

$$kH = \frac{1}{0.67} \frac{(1.15)(0.03)}{\pi} = 1.64 \times 10^{-2} \text{ m}^2/\text{s}$$

and

$$\log R_0 = \frac{1.87}{0.67} = 2.79$$

or

$$R_0 = 610 \text{ m}$$

Since $r \ll 0.1R$ for the last 2 holes, too much reliance should not be placed on these values in drawing the s/r relationship. Remember that if Q is measured in m^3/s then kH is in m^2/s.

5.6.3 Test pumping a well in an unconfined aquifer. It is assumed there is no supply to the aquifer from above (rainfall) or below. From equation 5.9

$$H^2 - h^2 = \frac{Q_0}{\pi k} \ln \frac{R}{r}$$

The left-hand side of this equation can be written as $(H - h)(H + h)$. But since $(H - h) = s$ and hence $h = (H - s)$, then the left-hand side can also be written as $s(2H - s)$. Therefore

$$s = \frac{Q_0}{\pi k (2H - s)} \ln \frac{R}{r}$$

or

$$s = \frac{Q_0}{2\pi k (H - \frac{s}{2})} \ln \frac{R_0}{r} \qquad\qquad (5.15)$$

for points in the vicinity of the well in a large aquifer.

When the drawdown is small in comparison to the depth of the aquifer, $s/2$ is negligible and the drawdown formula is the same as for a confined aquifer (equation 5.14).

Assuming that H is known approximately from the drilling of observation wells and $s/2$ is not negligible, then rewriting equation 5.15 gives

$$s' = s \left(1 - \frac{s}{2H} \right) = \frac{Q_0}{2\pi kH} \ln \frac{R_0}{r}$$

and s' can be plotted against r in the manner of figure 5.17.

Example 5.6 Figure 5.18 is a plot of observed drawdown against *r* for a well in an unconfined aquifer discharging a constant 0.03 m³/s. The depth of the aquifer below the phreatic surface has been established during drilling as about 20 m. Determine the formation constants.

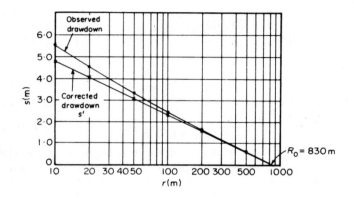

Figure 5.18 *Observed and corrected drawdown s' for an unconfined aquifer*

s', the corrected drawdown, is computed for each observed drawdown and from the straightline relationship with r

$$s' = A + B \log r$$

As $s' = 4.8$ when $r = 10$, hence $\log r = 1.00$
and as $s' = 0$ when $r = 830$, hence $\log r = 2.92$

Therefore

$$A - 2.92B = 0$$
$$A - 1.00B = 4.8$$

Hence

$$A = 7.30 \text{ and } B = 2.50$$

Also

$$kH = \frac{1}{2.50} \cdot \frac{(0.03)1.15}{\pi} = 4.39(10^{-3}) \text{ m}^2/\text{s}$$

so

$$R_0 = 830 \text{ m}$$

References

1. CRUSE, K. A review of water-well drilling methods. *J. Eng. Geol.*, **12** (1979) 63
2. STONER, R. F. *et al.* Economic design of wells. *J. Eng. Geol.*, **12** (1979) 63
3. HUISMAN, L. *Groundwater Recovery*, Macmillan, London, 1972
4. BRERETON, N. R. Step-drawdown pumping tests for the determination of aquifer and bore-hole characteristics. *Water Resources Council Tech. Report 103*, Washington D.C., January 1979
5. VERRUIT, A. *Theory of Groundwater Flow*, 2nd edition, Macmillan, London, 1981

Further reading

ARONOVICI, V. S. The mechanical analysis as an index of subsoil permeability. *Proc. Am. Soc. Soil Sci.*, **11** (1947) 137

CEDERGREEN, H. R. *Seepage, Drainage and Flow Nets*, John Wiley, New York, 1967

CHILDS, E. C. and COLLIS-GEORGE, N. The permeability of porous materials. *Proc. Roy. Soc.*, **A201** (1950) 392

KIRKHAM, D. Measurement of the hydraulic conductivity of soil in place. *Symposium on Permeability of Soils. American Society for Testing and Materials, Special Tech. Publication 163*, 1955, p. 80

RUSHTON, K. R. and REDSHAW, S. C. *Seepage and Groundwater Flow*, Wiley, 1979

TODD, D. K. *Ground Water Hydrology*, John Wiley, New York, 2nd edition, 1980

WENZEL, L. K. Methods for determining permeability of water bearing materials. *U.S. Geological Survey Water Supply Paper 887*, Washington D.C., 1942

Problems

5.1 Rainfall of 2.50 m per annum falls on a strip of land 1 km wide lying between two parallel canals, one of which (canal A) is 3 m higher than the other (canal B). The infiltration rate is 80 per cent of the rainfall and there is no run-off. The aquifer that contains the canals is 10 m deep below the level of canal B and both canals penetrate it fully. It is underlain by a horizontal impermeable stratum. Compute the discharge per metre length into both canals, assuming their boundaries are vertical and the aquifer coefficient of permeability $K = 10$ m/day.

5.2 A fully-penetrating well of 0.5 m diameter in a confined aquifer has been test-pumped at a rate of 0.025 m^3/s until steady-state conditions have been reached.

Observation wells at various distances from the well show the following results:

Distance r from well (m)	20	50	200	500
Drawdown s (m)	0.79	0.63	0.39	0.235

Using these results, determine the formation constants for the aquifer and, from them, the maximum constant yield of the well if the drawdown in the well is not to exceed 3 m. (Allow 0.5 m for screen and pack losses.)

6 Surface Runoff

6.1 The engineering problem

Rainfall, if it is not intercepted by vegetation or by artificial surfaces such as roofs or pavements, falls on the earth and either evaporates, infiltrates or lies in depression storage. When the losses arising in these ways are all provided for, there may remain a surplus that, obeying the gravitation laws, flows over the surface to the nearest stream channel. The streams coalesce into rivers and the rivers find their way down to the sea. When the rain is particularly intense or prolonged, or both, the surplus *runoff* becomes large and the stream and river channels cannot accept all the water suddenly arriving. They become filled and overflow and in so doing they do great harm to the activities of men. The most serious effect of flooding may be the washing away of the fertile top soil in which crops are grown, and of which there is already a scarcity on the earth. In urban areas there is great damage to property, pollution of water supplies, danger to life and often total disruption of communications. In agrarian societies floods are feared like pestilence because they can destroy crops, cattle and habitations, and bring famine in their wake.

The hydraulic engineer, in dealing with runoff, has to try to provide answers to many questions, of which some of the more obvious are

   (i) how often will floods occur?
   (ii) how large will they be and to what level will they rise?
   (iii) how often will there be droughts?
   (iv) how long may these droughts continue?

Questions of this kind are all variations of one, which is concerned with the magnitude and duration of runoff from a particular catchment with respect to time. They can be resolved best by the determination of the frequency and duration of particular discharges from observations over a long period of time, though if such observations are not available, estimations can still be made at various probabilities.

A second group of questions arises in using the curves of runoff frequency and duration once found; for example

    (i) how can the volume of discharge be reduced?

    (ii) how can the cost of flood prevention be compared with the damage that will arise if no measures are taken?

    (iii) how valuable is stored flood water in times of drought?

These questions are not directly related and each involves a quite different and distinct approach, though the same techniques can be used in answering more than one. In this and the following sections ways in which some of these questions can be answered will be sought.

6.2 Catchment characteristics and their effects on runoff

It is appropriate to consider how various properties of the *catchment area* affect the rate and quantity of discharge from it. By 'catchment area' is meant the whole of the land and water surface area contributing to the discharge at a particular stream or river cross-section, from which it is clear that every point on a stream channel has a unique catchment of its own, the size of catchment increasing as the control point moves downstream, reaching its maximum size when the control is at the sea coast.

There are many catchment properties that influence runoff and each can be present to a large or small degree. The intention in analysing them separately is to try to determine the effect of each characteristic on precipitation and its subsequent drainage from the catchment through the river channels.

(a) *Catchment area.* The area as defined at the beginning of this section is usually, but not necessarily, bounded by the topographic *water-divide*. Figure 6.1 shows a hypothetical cross-section through the topographic water-divide of a catchment. Because of the underlying geology it is perfectly possible for areas beyond the divide to contribute to the catchment. The true boundary is indeterminate, however, because although some of the groundwater on the left of the divide in the figure may arrive in catchment B, the surface runoff will stay in catchment A. Here the infiltration capacity of the soil and the intensity of the rainfall will influence the portion of the rainfall that each catchment will collect.

If runoff is expressed not as a total quantity for a catchment but as a quantity per unit area (usually m^3/s per square kilometre or $ft.^3/s$ per square mile), it is observed, other things being equal, that peak runoff decreases as the catchment area increases. This is due to the finite time taken by the water to flow through the stream channels to the control section (the *time of concentration*) and also to the lower average intensity of rainfall as storm size increases (see section 2.8.4.). Similarly, minimum runoff per unit area is increased because of the greater areal extent of the groundwater aquifers and minor local rainfall.

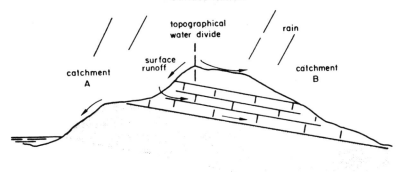

Figure 6.1 *Watershed defined by geology as well as topography*

(b) *Main stream length (MSL)*. This is measured in km from the gauging station or catchment outfall. When measuring from maps it is usual to use a standard routine to remove subjectivity (for example, set dividers at 0.1 km on a 1:25 000 scale map).

(c) *Slope of catchment*. The more steeply the ground surface is sloping the more rapidly will surface runoff travel, so that concentration times will be shorter and flood peaks higher. Infiltration capacities tend to be lower as slopes get steeper, since vegetation is less dense and soil more easily eroded, thus accentuating runoff.

Slope can be enumerated by covering a catchment contour map with a rectilinear grid and evaluating the slope, perpendicular to the contour lines at each grid point as shown in figure 6.2a. A frequency distribution of these numbers can then be plotted as in figure 6.2b. Different catchments can be compared on the same plot, the relatively steep frequency curves indicating catchments of fast runoff and flat curves the converse.

A simpler method is to describe the slope in m/km between two points on the main stream. The standard points used in Britain are 10 per cent and 85 per cent of the main stream length above the point of interest. Slope defined in this way is written as S1085.

(d) *Catchment orientation*. Orientation is important with respect to the meteorology of the area in which the catchment lies. If the prevailing winds and lines of storm movement have a particular seasonal pattern, as they usually have, the runoff hydrograph will depend to some degree on the catchment's orientation within the pattern. The grid of figure 6.2a can be used for this feature also, by the evaluation of the angle between slope direction and the north–south

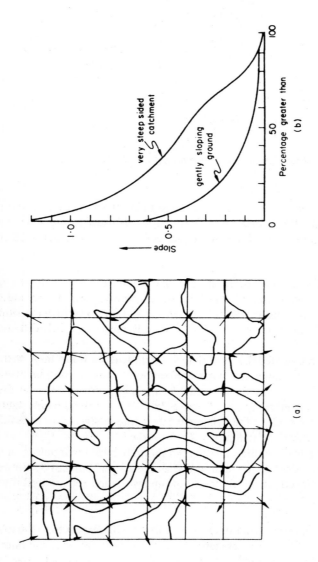

Figure 6.2 (a) Rectilinear grid to evaluate catchment slope and orientation. (b) Frequency curves for comparison of catchment steepness

meridian (say) at each grid point and the subsequent plotting of a circular frequency diagram like that of figure 6.3, similar to a wind rose.

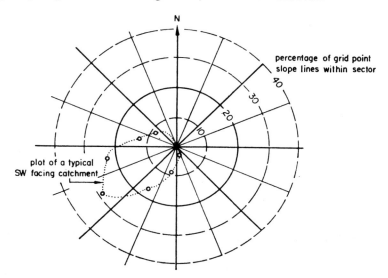

Figure 6.3 *Orientation diagram*

(e) *Shape of catchment.* The effect of shape can best be demonstrated by considering the hydrographs of discharge from three differently shaped catchments of the same area, shown in figure 6.4, subjected to rainfall of the same intensity. If each catchment is divided into concentric segments, which may be assumed to have all points within an equal distance along the stream channels from the control point, it can be seen that the shape A will require 10 time units (say hours) to pass before every point on the catchment is contributing to the discharge. Similarly B will require 5 h and C $8\frac{1}{2}$ h. The resulting hydrographs of runoff will be similar to those shown in figure 6.4, each marked with the corresponding lower-case italic letter. B gives a faster stream rise than C and A, and similarly a faster fall, because of the shorter travel times.

This factor of shape also affects the runoff when a rainstorm does not cover the whole catchment at once but moves over it from one end to the other. For example, consider catchment A to be slowly covered by a storm moving upstream that just covers the catchment after 5 h. The last segment's flood contribution will not arrive at the control for 15 h from commencement, so that the hydrograph a of figure 6.4 will now have the form of curve a_1 on that figure. Alternatively if the storm were moving at the same rate downstream, the flood contribution of time-segment 10 would arrive at the control point only 5 h after that of segment 1, so that a rapid flood rise (a_2 in figure 6.4) would occur. The effect of changing the direction of storm movement on the semicircular and fan-shaped catchments will be less marked than this but still appreciable.

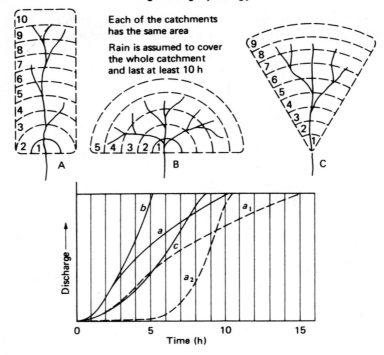

Figure 6.4 *The effect of shape on catchment runoff*

(f) *Annual average rainfall.* The standard annual average rainfall (SAAR) for the period 1941–1970 for the British Isles is given in the ten section maps for Britain and four for Ireland in appendix A.

(g) *Stream frequency.* The pattern of stream development in a catchment can have a marked effect on the runoff rate. A well-drained catchment will have comparatively short concentration times and hence steeper flood-rise hydrographs than a catchment with many surface depressions, marshy ground and minor lakes for example. One way is to measure the *stream density* of the catchment, by measuring the length of stream channel per unit area: another way (STMFRQ) used in the Flood Studies Report is to express it by the number of stream junctions per unit area, using standard scale maps.

(h) *Baseflow index (BFI).* This is an index calculated as the ratio of the flow under the separated hydrograph (see section 7.3) to the flow under the total hydrograph. It is a good indicator of the catchment's underground storage, which is dependent on the solid geology. Methods of estimation are given in section 6.6.4 and figure 6.5 illustrates how BFI is calculated from flow data.

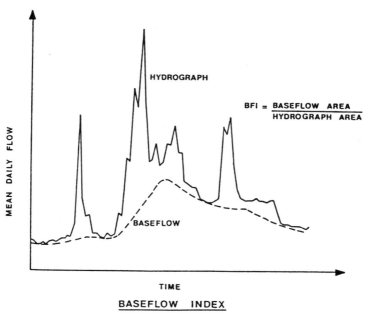

Figure 6.5 *Calculation of BFI from flow data*

(i) *Lake and reservoir area.* These act as surface water stores and have the effect of smoothing out the hydrographs of catchments that contain them.

(j) *Soil-moisture deficit.* This is a climatic parameter that is dependent on rainfall and evapotranspiration and has been discussed in chapter 4.

There are other factors in addition to these, including *altitude* (with its effect on temperature and the presence of snow in winter), *land use* (whether forested or arable grassland), the proportion of *urban development* and the *condition of stream channels.*

6.3 Climatic factors

In section 6.2(e) reference has already been made to the effect of storm movement on surface runoff. If the areal extent of a storm is such that it does not cover the whole of a catchment, the runoff will be less than from complete coverage.

The form of precipitation also has an influence, since snowfall and freezing temperatures can effectively put the expected runoff into storage and reduce evapotranspiration.

The main effect of climate however is in rainfall *intensity* and *duration.* Rainfall intensity has a direct bearing on runoff since once the infiltration capacity is exceeded all the excess rain is available and flows to the surface water-courses. Intensities vary greatly, the maximum usually occurring in severe local thunderstorms. It will be realised that severe local maxima are recorded only fortuitously, so that the highest recorded intensities will certainly have been exceeded many times.

Since intensity represents, depth/time, it cannot be considered separately from duration. The same depth of rainfall delivered over two different durations will produce different runoff rates. Also, different climates will produce different meteorological conditions leading to different types of rain, which may inherently have quite dissimilar durations. For example, in England a thunderstorm may conceivably produce rainfall intensities as great as 10 mm/min (see figure 2.4) but it cannot be imagined that such storms can last for more than a period of minutes, whereas the *monsoon* rain in India can fall continuously for weeks at average intensities greater than 10 mm/h, a condition never approached in most other parts of the world.

The influence of duration on the hydrograph of runoff can be seen from figure 6.6, where a *uniform-intensity storm* causes the hydrograph of streamrise a. Such storms may be defined as covering the whole catchment area, over which the depth of rainfall is reasonably constant and delivered at a constant rate. Although rare in nature, they are used in hydrology to determine characteristics of catchments. After a certain time, t_c, the period of concentration, the rate of runoff becomes constant. A hydrograph of this form is typical of only very small catchments; for example, paved urban areas, where such constant runoff is quickly achieved. Natural catchments of any size have periods of

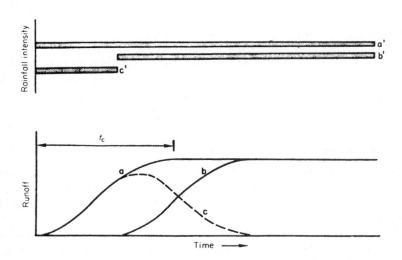

Figure 6.6 *Hydrograph of short-period storm runoff derived from two long-duration uniform-intensity storms*

concentration longer, as a general rule, than periods of uniform-intensity rain. The effect of short periods of rain can be found by subtracting the hydrographs of two long periods, exceeding t_c, and separated in time by the short period, one from the other. In figure 6.6, a and b are the result of rainfalls a' and b' respectively. Their subtraction leaves the short-period rain c' and its resulting hydrograph c, which is the typical shape of most natural hydrographs.

6.4 Rainfall/runoff correlation

While there is a general cause-and-effect relationship between rainfall and the resulting runoff, it will be clear by now that it is not a direct one. By the time that evaporation, interception, depression storage, infiltration and soil-moisture deficiency are taken into account and the resulting residual rainfall at various intensities applied to catchments that vary in size, slope, shape, altitude, subsurface geology and climate, the relationship must include extreme values that defy rational correlation, at least in the short term.

Notwithstanding the foregoing, it may be possible to establish an empirical relationship for a particular catchment based on annual precipitation and runoff. To do this it is best to use a *water-year* rather than a calendar year. This is a 12-month period starting and finishing at the time of seasonal minimum flow. If precipitation and runoff are plotted against each other as catchment depths, a correlation like that shown in figure 6.7, may be obtainable. In temperate

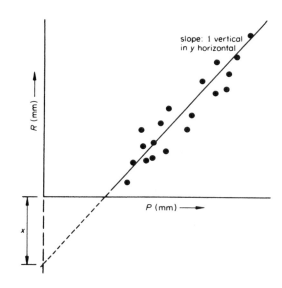

Figure 6.7 *Rainfall/runoff correlation*

and tropic, humid climates such a straight-line relationship generally is found, where if P is annual precipitation, annual runoff R is expressed as

$$R = \frac{P}{y} - x \qquad (6.1)$$

so that the annual rainfall can be used to obtain a first approximation to the annual runoff. Young has analysed the correlation on a world-wide basis [1].

Variations from the straight line may be due to conditions in the preceding year that gave markedly higher or lower groundwater levels, or to variations in the seasonal distribution of rainfall. The method can be used also for wet months in humid climates when the ground is saturated, but beyond such narrow limits it is not valid.

Although the application of a relationship like equation 6.1 is restricted, it can nevertheless be a useful method for estimating total annual runoff or completely ungauged catchments, if they are in similar climates and of similar size and character. The relationship has been developed for use with synthetic hydrographs, using catchment wetness index and soil characteristics as parameters [2].

Further refinements are possible taking into account the particular period of the year, the antecedent precipitation index (see section 4.4.2) and the storm duration as well as depth, so that relationships may be derived for particular storms on a particular catchment. Coaxial graphs may be produced which take the various variables into account. A relationship of this form is shown in figure 4.7.

6.5 Flow rating curves: their determination, adjustment and extension

6.5.1 Definition. A rating curve is a graph that shows the connection between the water level elevation, or *stage* of a river channel at a certain cross-section, and the corresponding discharge at that section. A typical rating curve is shown in figure 6.8. It can be seen that the curve is drawn through a cluster of points, each of which represents the results of a river discharge measurement. Such measurements can be made in a number of ways, of which the most important are

(1) velocity-area methods,
(2) flow-measuring structures,
(3) dilution gauging.

6.5.2 Velocity–area methods. These are conventional for medium to large rivers and involve the use of a *current meter*, which is a fluid velocity-measuring instrument. A small propeller rotates about a horizontal shaft, which is kept

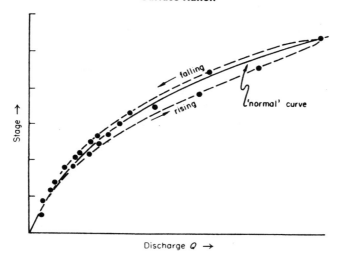

Figure 6.8 *Flow rating curve*

parallel to the streamlines by tail fins. The instrument is ballasted to keep it as nearly as possible directly below the observer. Another version of the instrument has a circlet of small conical cups disposed horizontally about the suspension axis.

Each revolution of the propeller is recorded electrically through a cable to the observer and the number of revolutions is counted by the observer or automatically over a short period (say 1 or 2 minutes). These observations are converted into water velocities from a calibration curve for the instrument. By moving the meter vertically and horizontally to a series of positions whose co-ordinates in the cross-section are known, a complete velocity map of the section can be drawn and the discharge through the cross-section calculated. Figure 6.9 shows a modern current meter assembled for use on its supporting cable; this meter can also be used as a depth measure.

Observations are made by lowering the meter from a bridge, though if the bridge is not a single-span one there will be divergence and convergence of the streamlines caused by the piers, which can cause considerable errors. In many instances, however, the gauging site, which should be in as straight and uniform a reach of the river as is possible, will have no bridge, and if it is deep and in flood a cable to hold some stable boat must be provided, together with a lighter measuring cable to determine horizontal position in the cross-section.

Since the drag on a boat with at least two occupants and suspended current meter is considerable, a securely fastened steel cable should be used. The presence of suitable large trees at a particular site often necessitates its choice for this reason. Alternatively, cableways are sometimes used to suspend the meter, either

Figure 6.9 *A modern helix current meter by Hilger and Watts Ltd*

from a manned cable-car or directly from the cable, the instrument in this case being positioned by auxiliary cables from the river banks.

Depths should always be measured at the time of velocity observation since a profile can change appreciably during flood discharges. Observers should also remember such elementary rules as to observe the stage before and after the discharge measurement, and to observe the water slope by accurate levelling to pegs at the river level as far upstream and downstream of the gauging site as is practicable, up to say 500 m in each direction.

As water velocities increase in high floods the ballasted current meter will be increasingly swept downstream on an inclined cable. The position of a meter in these circumstances can be found reasonably accurately if the cable angle is measured. Ballast can be increased but only within limits. Rods can be used to suspend the meter but a rigid structure in the boat will then be required to handle the rods, calling for a stable platform of the catamaran type. Rod vibration and bending are common in deep rivers unless diameters exceed 50 mm, by which time the whole apparatus is getting very heavy and unmanageable.

It will be appreciated that since each river is unique, each will require a careful assessment of its width, depth, likely flood velocities, cable support facilities, availability of bridges, boats, etc. before a discharge measurement programme is started.

From many observations on many rivers it has been established that the variation of velocity integrated over the full depth of a river can be closely approximated by the mean of two observations made at 0.2 and 0.8 of the depth. If time and circumstances preclude even two observations at each horizontal intercept then one reading at 0.6 depth measured from the surface will approximate the average over the whole depth.

The discharge at the cross-section is best obtained by plotting each velocity observation on a cross-section of the gauging site with an exaggerated vertical scale. *Isovels* or contours of equal velocity are then drawn and the included areas measured by planimeter. A typical cross-section, so treated, is shown in figure 6.10a. Alternatively the river may be subdivided vertically into sections and the mean velocity of each section applied to its area (figure 6.10b). In this method, the cross-sectional area of any one section should not exceed 10 per cent of the total cross-sectional area.

A check should always be made using the slope-area method of section 6.5.7 (iii) and a value obtained for Manning's n. In this way a knowledge of the n values of the river at various stages will be built up, which may prove most valuable in extending the discharge rating curve subsequently.

To ensure uniformity in the techniques of current-meter gauging the International Organisation for Standardisation (ISO) has published various recommendations and, in the United Kingdom, BS 3680 'Measurement of liquid flow in open channels' refers [3]. In the USA, the U.S. Geological Survey and the U.S. Bureau of Reclamation have established practice [4–7].

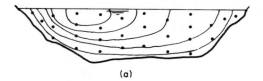

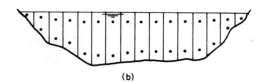

Figure 6.10 *Two computational methods for discharge measurement:
(a) isovel plotting of cross-sectional discharge; (b) sectional averaging
of observations at 0.2 depth and 0.8 depth*

6.5.3 Flow-measuring structures. These are designed so that stream discharge is
made to behave according to certain well-known hydraulic laws. For example,
the discharge per unit length over a weir is a function of the head over the weir.
Many specialised weirs, such as V-notch, compound and Crump weirs, have been
designed to provide accurate discharge data by observations of water surface
level upstream of the weir. Flumes can similarly be used, where a stream is
channelled through a particular geometrically shaped regular channel section for
some distance before entering a length of different cross-section, usually made so
by side contractions or steps in the bed. Trapezoidal shapes are often used, and
narrow vertical sections have been employed for catchment discharge measurement in Wales. Figure 6.11 shows a flume of the latter type on the Plynlimon
experimental catchments of the Institute of Hydrology. Generally, however,
flow-measuring structures are confined to streams and fairly small rivers, since
for large flows and wide rivers they become extremely expensive to build [8-15].

BS 3680, Part 4 Weirs and Flumes, provides detailed dimensions for standard,
calibrated weirs and flumes of most types.

6.5.4 Dilution gauging. Dilution gauging is particularly suited to small turbulent
streams where depths and flows are inappropriate for current-metering and flow-
structures would be unnecessarily expensive and/or permanent.

Figure 6.11 *Narrow flow-measurement flume—Plynlimon catchments*

The method involves the injection of a chemical into the stream and the sampling of the water some distance downstream after complete mixing of the chemical in the water has occurred. The chemical can either be added by *constant-rate injection* until the sampling downstream reveals a constant concentration level, or administered in a single dose as quickly as possible, known as *gulp injection*. In this case samples over a period of time disclose the concentration–time correlation. In both cases the concentration of chemical in the samples is used to compute the dilution, and hence the discharge of the stream can be obtained. Figure 6.12 shows constant-rate injection of sodium dichromate from a Mariotte bottle (a constant-head device) in a mountain stream.

Analysis of the samples is by an automated colorimetric procedure that estimates the concentration of very small amounts of the chromium compound by comparison with a sample of the injection solution. The equipment is expensive and specialised. References [16, 17] give comprehensive guidance.

Another method, developed by Littlewood [18], deserves mention at greater length as the required equipment is simple and relatively cheap. The method depends on the electrical conductivity of solutions of common salt (NaCl) in the stream water and is a version of the relative-dilution-gauging method of Aastad and Søgnen [19].

Figure 6.12 *Dilution gauging: dispensing sodium dichromate solution from a Mariotte bottle*

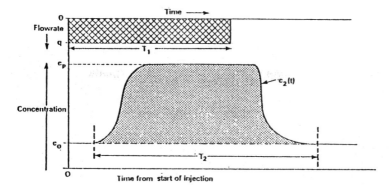

Figure 6.13 *Extended constant-rate injection method* [18]

Referring to figure 6.13, it can be shown that if

Q = streamflow (l/s)
V = volume of injection solution of concentration c_1 (l)
T_1 = duration of injection (s)
q = injection rate (l/s)
T_2 = duration of solute wave
c_1 = concentration of injection solution (mg/l)
c_2 = variable concentration of the streamwater
c_0 = the background concentration (mg/l)

then

$$qc_1 T_1 = Q(\overline{c}_2 - c_0)T_2 \text{ (where } \overline{c}_2 \text{ is the mean of all } c_2 \text{ values)}$$

which may be written

$$Vc_1 = Q(\overline{c}_2 - c_0)T_2$$

Hence

$$Q = \frac{V}{T_2} \times \frac{c_1}{(\overline{c}_2 - c_0)} \tag{6.2}$$

and

$$Q = \frac{V}{T_2} \times \overline{N} \tag{6.3}$$

where $\overline{N}$ is the mean dilution ratio.

It is not necessary to measure the shaded area under the curve of figure 6.13 since, provided the curve is defined by closely spaced ordinates, the average of these (in the shaded area) is all that is needed. This holds good for any length of injection time, including a 'gulp', and not just for the constant-rate injection of figure 6.13 which is simply a special case of a general method.

It is necessary now to move from solution concentration to conductivity but two problems must be overcome:

(a) the relationship between concentration of salt and conductivity of the solution is non-linear
(b) natural streams have varying background conductivities.

The first of these is overcome by constructing a graph of conductivity against concentration which for very weak solutions is virtually linear. This is so for solvents of initially different conductivities, where the gradients of the lines in figure 6.14 are practically parallel.

It follows that, for changes of concentration, Δc, and conductivity, $\Delta c'$, in the linear range of weak solutions

$$\Delta c = K_1 \Delta c'$$

K_1 for NaCl in water is approximately 0.51 μs cm^{-1}/mg l^{-1}.
Equation 6.2 may now be written

$$Q = \frac{V}{T_2} \times \frac{c_1}{K_1 \overline{\Delta c'}}$$

where $\overline{\Delta c'}$ is the average of $(c_2' - c_0')$.   Now from equation 6.3

$$\overline{N} = \frac{c_1}{(\overline{c_2} - c_0)} = \frac{c_1}{K_1 \overline{\Delta c'}}$$

which can be written

$$\overline{N} = K_2 \frac{c_1'}{\overline{\Delta c'}} \tag{6.4}$$

where K_2 is a combination of K_1 and a multiplier for converting a particular injection solution conductivity, c_1', to concentration, c_1.

If a small volume, v, of the particular strong solution is added to a larger volume V^* of the streamwater, giving rise to a dilution ratio N^*, and the change in conductivity is recorded as $\Delta c'^*$, then we can write

$$N^* = K_2 \frac{c_1'}{\Delta c'^*} \tag{6.5}$$

where $N^* = \dfrac{V^*}{v}$.

Now dividing equation 6.4 by equation 6.5 gives

$$\frac{\overline{N}}{N^*} = \frac{\Delta c'^*}{\overline{\Delta c'}}$$

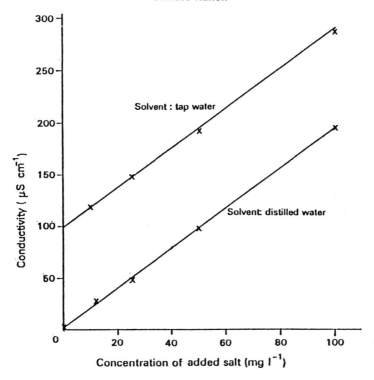

Figure 6.14 *Conductivity against concentration of added salt (weak solution)* [18]

from which it follows that

$$Q = \frac{V}{T_2} \times \frac{V^*}{v} \times \frac{\overline{\Delta c'}^*}{\overline{\Delta c'}}$$

This means that the discharge may be measured by pouring a known volume (V) of a strong salt-solution into the stream, measuring the change in conductivity of the water at the downstream end of the mixing length (to find $\overline{\Delta c'}$) over time T_2 and measuring the change in conductivity of the weak solution obtained by controlled dilution of the strong solution in a separate operation.

The method is independent of the make and calibration of the conductivity meter. It is not necessary to know the concentration of the strong solution and very simple equipment is needed.

Full details of the method, comparative test results and typical quantities and dimensions are given in reference [18].

6.5.5 Ultrasonic gauging. Among recent developments in river gauging is the method of ultrasonics. It can provide continuous unattended discharge measurement or, more conventionally, at some specified regular interval. The method depends on the transmission of ultrasonic pulses between two sets of transmitter/receivers situated on either bank of the river, and so arranged so that the line of transmission is at 45°, or thereabouts, to the flow direction. Since the speed of transmission in one direction is greater than in the other, owing to the component of the water flow velocity along the line of transmission, the pulse travel times are different. This difference is a function of the mean water velocity at the pulse level.

The usual arrangement consists of a section of a river with a rectangular lined cross-section at least as long as its width, with pairs of transmitters at various levels and a water level recorder. Automatic samples and recording can then provide a comprehensive discharge record at the station to an accuracy at least as good as any other method.

For detailed descriptions of installations, methods and analyses of results, interested readers should consult references [20, 21].

The methods of sections 6.5.2, 6.5.3, 6.5.4 and 6.5.5, used singly or in conjunction, will establish the correlation for any stream or river discharge with stage.

A rating curve when established in this way, enables a single observation of stage, made each day at a set time, by an unskilled observer, to be converted into a discharge rate and hence into a finite quantity of water, flowing at the observation point. The difficulty about rating curves is to obtain enough points at times of high discharge to enable an accurate correlation to be obtained.

6.5.6 Rating-curve adjustment. So far rating curves have been discussed in terms that imply they are simply median lines drawn through a scatter of observation points. This is not the case. If each discharge point has been noted as being measured on a 'falling' or 'rising' stage, the curve would strictly speaking become a loop as shown by the dashed line in figure 6.8.

This variation, or looping effect is due to several causes. The first of these is *channel storage*. As the surface elevation of a river rises, water is temporarily stored in the river channel.

Example 6.1. Suppose that the gauge shows a rise at the rate of 0.2 m/h during a discharge measurement of 100 m³/s and the channel is such that this rate of rise may be assumed to apply to a 1000 m reach of river between the measurement site and the reach control (*the control of a river reach is the section at which the profile of the river changes*).

Let the average width of channel in the reach be 100 m. Then the rate of change of storage in the reach, dS is given by

$$dS = 1000 \times 100 \times 0.2$$
$$= 20\,000 \text{ m}^3/\text{h}$$
$$= 5.6 \text{ m}^3/\text{s}.$$

The discharge measurement should be plotted on the rating curve as 94.4 m³/s (not as 100), since this is the discharge past the *control* corresponding to the mean gauge height.

The second reason for the looping of rating curves is the variation in surface slope that occurs as a flood wave moves along the channel. Figure 6.15 represents

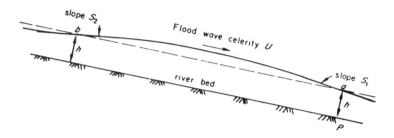

Figure 6.15 *Flood wave slope variation*

a longitudinal section of a flood wave passing along a river channel. As point a passes the gauging site, the gauge reads h, the river cross-section is A and the slope of the water surface S_1. When the flood wave has moved on so that point b is at the gauge site, the gauge reading h and the cross-section A are the same but the slope is now S_2. From the Manning formula

$$Q = Av = \frac{AR^{\frac{2}{3}}S^{\frac{1}{2}}}{n}$$

and so two different discharges will occur in the two cases since S changes while A, R and n remain constant.

Since the rising stage is associated with the greater slope discharge, measurements taken on a rising stage will plot to the right on the rating curve of figure 6.8 and those on a falling stage, to the left. Depending on the 'peakiness' of the flood wave, it often happens that maximum discharge occurs before the maximum stage is reached, since the influence of the steeper slope on velocity may outweigh the slight increase in cross-section area.

It is generally necessary to correct the discharge measurements taken on either side of flood waves to the theoretical steady-state condition, because the great majority of gauge readings are taken by unskilled observers once a day without any indication of whether the stage is rising or falling. By using a corrected, or steady-state, curve, the rising and falling stage observations will

balance in the long run, and no value judgement or second daily visit to the gauge will be required from the observer. The correction can be made as follows.

From the Manning equation, the steady-state discharge Q in a channel of a given roughness and cross-section, is given as

$$Q \propto \sqrt{S} \tag{6.6}$$

where S = steady-state slope.

When the slope is not equal to S, as is the case in conditions of rising or falling stage, the actual discharge Q_a is given by

$$Q_a \propto \sqrt{(S \pm \Delta S)} \tag{6.7}$$

Referring now to figure 6.16, ΔS can be expressed in terms of the rate of change in stage and flood wave celerity, U.

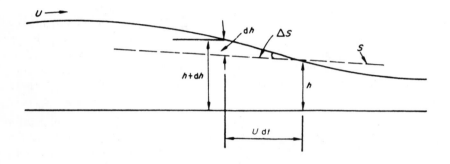

Figure 6.16 *Change in slope of a flood wave with time*

The figure represents an advancing flood wave and rising state. The gauge reading at the commencement of the discharge measurement is h and at its conclusion, dt, later, $h + dh$. In this time the wave has advanced Udt, and

$$\Delta S = \frac{dh}{Udt} = \frac{dh/dt}{U} \tag{6.8}$$

dh/dt being positive for a rising stage, negative for a falling stage. Combining equations 6.6, 6.7 and 6.8

$$\frac{Q_a}{Q} = \sqrt{\left(\frac{S + (dh/dt)/U}{S}\right)}$$

or

$$\frac{Q_a}{Q} = \sqrt{\left(1 + \frac{dh/dt}{US}\right)} \tag{6.9}$$

If the discharge measurement is taken at a site with two gauges, one upstream and one downstream, in the same reach, then all the terms in equation 6.9 are measured except Q and U, Q being the steady-state discharge required and U the flood wave celerity. There are now several ways to proceed. The first is to take an empirical value for U. Corbett *et al.* [4] suggest that in a fairly uniform channel, in flood conditions, the celerity U is approximately equal to 1.3 times the mean water velocity, or

$$U = 1.3 \, \frac{Q_a}{A}$$

where A = cross-sectional area.

From which

$$Q = \frac{Q_a}{\sqrt{\left(1 + \frac{A \cdot dh/dt}{1.3 \, Q_a S}\right)}} \tag{6.10}$$

Example 6.2 A river discharge measurement made during a flood indicated $Q_a = 3160 \text{ m}^3/\text{s}$. During the measurement, which took 2 h, the gauge height increased from 50.40 to 50.52 m. Level readings taken at water surface 400 m upstream and 300 m downstream of the observation site differed by 100 mm. The river was 500 m wide with an average depth of 4 m at the time of measurement. At what co-ordinate should the measurement be plotted on the rating curve?

The cross-sectional area of the river, A = 500 m × 4 m
$$= 2000 \text{ m}^2$$
Therefore
$$\text{average water velocity} = \frac{Q_a}{A} = \frac{3160}{2000} = 1.58 \text{ m/s}$$

Assume the flood wave celerity $U = 1.3 \times 1.58 = 2.045 \text{ m/s}$: $dh/dt = 0.12 \text{ m}/7200 \text{ s} = 1.67 \times 10^{-5}$, and $S = 0.1/700 = 1.43 \times 10^{-4}$. Then for a rising river, from equation 6.5

$$Q_{corr} = \frac{3160}{\sqrt{\left(1 + \frac{1.67 \times 10^{-5}}{2.054 \times 1.43 \times 10^{-4}}\right)}} = \frac{3160}{\sqrt{(1.057)}} = 3074 \text{ m}^3/\text{s}$$

and taking the mean gauge height, the corrected co-ordinates are 50.46 m and 3074 m³/s.

An alternative procedure, due to Boyer [22] is available, where values for neither U nor S need be obtained. If a sufficient number of observations are available, including measurements taken during rising and falling stages and in steady states, then a rating curve can be drawn as a median line through the uncorrected values. The steady-state discharge Q can now be estimated from the median curve. Since Q_a and dh/dt are measured quantities and therefore known, equation 6.9 yields the term $1/US$ for each measurement of discharge.

The term $1/US$ is now plotted against stage and an 'average' curve fitted to the plotted points, as shown in figure 6.17. From the $1/US$ against stage relationship, new values of $1/US$ are obtained and inserted into equation 6.9 to yield the steady state Q. The new values of Q are then plotted against stage as the corrected, steady-state curve.

Another method, which uses the observed slope but avoids evaluating U, is due to Mitchell [23].

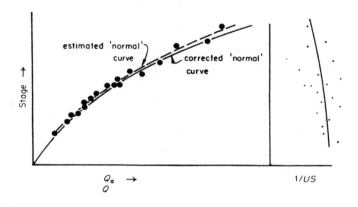

Figure 6.17 *Method of correcting discharge readings without computing U or S*

6.5.7 Extension of rating curves. As previously mentioned, the most difficult measurements of discharge to take are those at high flood, because of both the physical difficulties of high water velocity and floating debris, and the rare occurrence of the condition. It frequently happens, for example, that the condition for which river structures such as dams and bridges must be designed are defined as being 'such that occur no more often than once in 100 years'. This means that the structural designers want to know the probable discharge that will occur, on average, once every 100 years. This is sometimes referred to as 'the 100-year flood'.

If discharge measurements have been taken throughout the previous 100 years this design flood may not be too difficult to find. However, in the vast majority of cases, only stage readings will be available and then for limited periods only. If the hydraulic engineer has 30 years of continuous daily stage readings and discharge measurements that include even low flood conditions, then he is fortunate. Almost always he will have to extend the rating curve well beyond the last measured point, to estimate the discharge at particular stage levels recorded. Occasionally the stage levels reached in high flood are recorded only by lines of debris on the banks, or grasses caught in the branches of riparian trees and scrub. Such physical evidence is valuable.

There are a number of ways of attempting rating curve extension.

(i) *By fitting an equation to the curve.*

Usually an equation of the form $Q = k(h - a)^x$ is used, where

h = stage

k and x are constants derived from the observed portion of the curve, and

a = height in m (or ft.) between zero on the gauge and the elevation of zero flow.

Such a curve plots as a straight line on logarithmic paper and so can be easily extended. At best it is a questionable procedure since there is little theoretical justification for exponential laws operating and at high flows there may be quite abrupt changes in cross-section with rising stage.

(ii) *Stevens' method* [24].

This method is based on Chezy's formula

$$Q = AC\sqrt{(RS)}$$

where A = cross-sectional area

C = Chezy roughness coefficient

R = hydraulic radius

S = slope of the energy line.

If $C\sqrt{S}$ is assumed constant and D, the mean depth substituted for R then

$$Q = kA\sqrt{D}$$

Known values of $A\sqrt{D}$ and Q are plotted and often come close to a straight line which can be extended. Field values of $A\sqrt{D}$ above the measured rating can then be used from the extended line to plot Q against stage points on the rating curve.

The objection to this method is simply that $C\sqrt{S}$ is not a constant. However, as it takes account of the varying stream geometry it is a more rational procedure than (i).

Typical values of Chezy's C are given in appendix B.

(iii) *Slope-area method.* This method depends on hydraulic principles and presupposes that it is practical to drive in pegs or make other temporary elevation marks at the time of the flow measurement upstream and downstream of the discharge measuring site. These marks can subsequently be used to establish the water slope. Cross-sectional measurements will yield the area and hydraulic radius of the section. Then from Manning

$$Q = \frac{AR^{\frac{2}{3}}S^{\frac{1}{2}}}{n}$$

This method is sometimes criticised because of its dependence on the value of
n. Since n for natural streams is about 0.035 an error in n of 0.001 gives an error
in discharge of 3 per cent. This objection may be partially met by plotting n
against stage for all measured discharges, so that the choice of n for high stages
is not arbitrary but is taken from such a plot. If the high flood slope can be
measured then this method is probably the best one.

Typical values of Manning's n are given in appendix B.

It should be emphasised that all methods of rating curve extension are suspect
to some degree and should be resorted to only if flood measurements cannot be
obtained. The latter two methods above are both subject to errors that arise
from alteration of cross-section as a result of flood scour and subsequent low
water deposition; so cross-sections and mean depth measurements should be
taken as near to the time of discharge measurement as possible.

6.6 Volume and duration of runoff

6.6.1 Hydrographs. With an adjusted and well-measured rating curve, the daily
gauge readings can be converted directly to runoff volumes. A typical set of such
daily runoff figures is presented graphically in figure 6.18. Such a presentation is
called a hydrograph. Although figure 6.18 shows a hydrograph with a time base
of many months, hydrographs for smaller catchments can have time bases of
days or even hours.

While floods and droughts are important from many points of view, they
tend, as extremes, to be of comparatively short duration. For many water-
resource investigations it is equally important to know the total volumes of

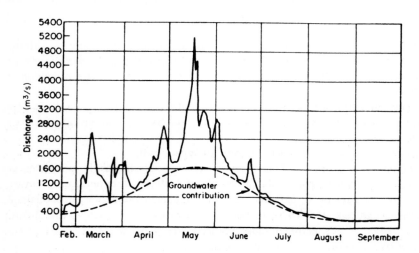

Figure 6.18 *Hydrograph of the river Euphrates at Hit, February to
September 1957 (after Directorate of Irrigation, Iraq)*

water that have to be dealt with over long periods of time; for example, in hydroelectric power generation the plant capacity must be chosen for some discharge well below the peak flood since otherwise much capacity would be almost permanently idle. For such purposes the most convenient means of presenting data are the *mass curve*, the runoff accumulation–time curve and the *flow duration curve*.

6.6.2 Mass curves. If the volumes denoted by the product of ordinate and time interval of a hydrograph are plotted against time by adding each new volume to the previous total, a cumulative mass curve of runoff is obtained. Such a curve is shown in figure 6.19.

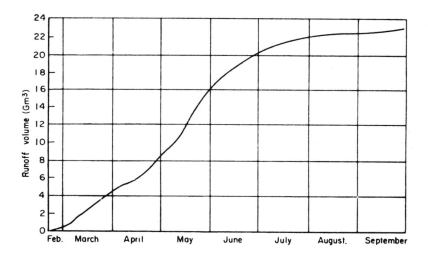

Figure 6.19 *Cumulative mass curve of runoff for the river Euphrates at Hit, February to September 1957*

Mass curves are extremely useful in reservoir design studies since they provide a ready means of determining storage capacity necessary for particular average rates of runoff and drawoff. Suppose for example that the mass curve *OA* of figure 6.20 represents the runoff from a catchment that is to be used for base load hydroelectric development. If the required constant drawoff is plotted on the same diagram, as line *OB*, then the required storage capacity to ensure this rate can be found by drawing the line *CD* parallel to *OB* from a point *C* at the beginning of the driest period recorded. The storage capacity necessary is denoted by the maximum ordinate *cd*. Normally much longer periods, as long as the record allows, are used for reservoir design and in many instances the drawoff is not constant nor continuous. In such cases, different techniques based on the same principles are used [25].

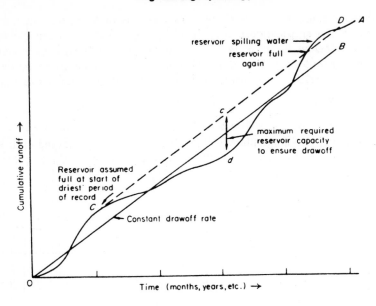

Figure 6.20 *The use of mass curves in reservoir design*

6.6.3 Runoff accumulation–time curves. Another way of using flow data in the design of reservoirs is to calculate reservoir filling times from daily flow readings. Since river flow varies seasonally, the time taken to accumulate particular volumes of runoff (equivalent to reservoir filling times) will vary with the time of year that filling commences, as well as with the stored volume required.

Engineers also need to know the probabilities of certain events taking place and the risks of not meeting particular objectives. These various criteria are met by *runoff accumulation–time frequency diagrams*. A typical example is shown in figure 6.21.

The derivation of such curves is straightforward but is too long to include here. A full description, with worked examples of the techniques required, is given in reference [26].

6.6.4 Flow duration curves. A flow duration curve (FDC) for a particular point on a river shows the proportion of time during which the discharge there equals or exceeds certain values. Such a curve is shown in figure 6.22. Flow duration curves for long periods of runoff are useful for deciding what proportion of flow should be used for particular purposes, since the area under a curve represents volume. Storage upstream of the gauging point in the forms of lakes or reservoirs will modify the FDC of a river that has been previously without such storage.

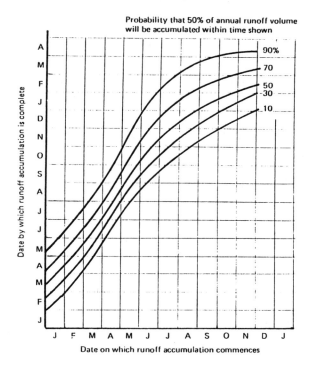

Figure 6.21 *Typical R.O. accumulation-time curve for a river* [26]

For many rivers the ratio of peak to minimum discharges may be two or more orders of magnitude and FDCs for points on them are often more conveniently drawn with the ordinate (Q) to a logarithmic scale and a normal probability scale used for the frequency axis. On such a graph, if the logarithms of the discharges are normally distributed, then the FDC plots as a straight line. This is often approximately the case.

Flow duration curves for different points on a river can be compared when presented in this more compact way by standardising them. The discharges are divided firstly by the area of contributing catchment and secondly by weighted annual rainfall over the catchment. The resulting discharges, in m^3/s or l/s, per unit area per unit annual rainfall, can then be compared directly. Figure 6.23 shows six FDCs for tributaries of the River Severn in England compared in this way. A composite regional standardised FDC for the upper reaches of the Severn and its tributaries was subsequently derived from figure 6.23 [27].

Such standardised curves provide a convenient means of synthesising an FDC for an ungauged point on a river, and for points on other rivers of similar geo-morphological nature in a similar climate. Obviously, care should be taken that these two conditions apply.

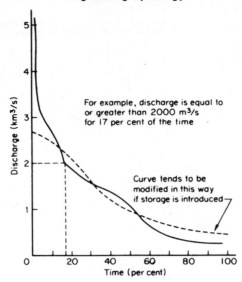

Figure 6.22   *Flow duration curve for the river Euphrates at Hit,
February to September, 1955 (derived from the data of figure 6.18)*

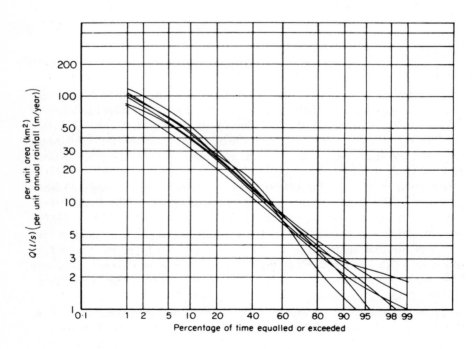

Figure 6.23 *Standardised FDCs for the Upper Severn*

Another way of standardising FDCs is to express Q in terms of Q/Q_m where Q_m is the mean flow. The use of a non-dimensional ordinate allows FDCs of all rivers, large and small, to be compared on the same graph. This provides a method proposed by the Institute of Hydrology for the synthesis of FDCs for ungauged catchments, involving the use of a Base Flow Index (*BFI*) [25]. The *BFI* is defined as the baseflow area/hydrograph area, i.e. the proportion of runoff that base flow comprises. *BFI* is effectively a function of the catchment's infiltration capacity and aquifer storage and so influences the shape of the FDC markedly. It may be determined for a gauged catchment as illustrated in figure 6.5, but for an ungauged one it is necessary to examine the catchment soils and geology in some detail. In the UK the FSR Winter Rain Acceptance Potential (WRAP) maps may be used since they are, in effect, soil-type maps. They are printed in appendix A using the notation RP. RP maps have five different *SOIL* types varying from

1. very high permeability – very low runoff to
2. very low permeability – very high runoff.

If the catchment is delineated on an RP map and the various proportions of it under each soil type are identified, then a general equation for *BFI* may be written

$$BFI = [0.6 + 0.23S_1 - 0.03S_2 - 0.12S_3 - 0.17S_4 - 0.20S_5] \qquad (6.11)$$

where S_1 to S_5 are the five SOIL categories.

A modification to equation 6.11 will be required for certain catchments which exhibit very high or very low baseflows. The modification depends on the use of 1/625,000 solid geology map of the British Isles to identify the underlying geology associated with extreme *BFI*s which may be as high as 0.9 in chalk areas and as low as 0.2 in carboniferous limestone and clay areas.

Having established the *BFI*, a standardised FDC can be selected from figure 6.24. This selected FDC now has its ordinates multiplied by the catchment Q_m to provide the particular FDC with numerical discharge values for the catchment.

The slope of the FDC gives an indication of the character of a river. A gentle slope indicates a river with few floods that is extensively supplied from groundwater, while a steeply sloping curve indicates a river with frequent floods and low flow periods, having little groundwater flow and being supplied mainly from runoff. Flow duration curves are usually produced from daily flow data but other time intervals can be chosen. For example, a D-day FDC is one that shows the proportion of D-day periods when the average discharge exceeds a certain value.

6.7 Estimation of mean flow, Q_m

The value of Q_m (also referred to as *ADF* or average daily flow) may be derived for a gauged catchment from the daily flow records by summation and averaging.

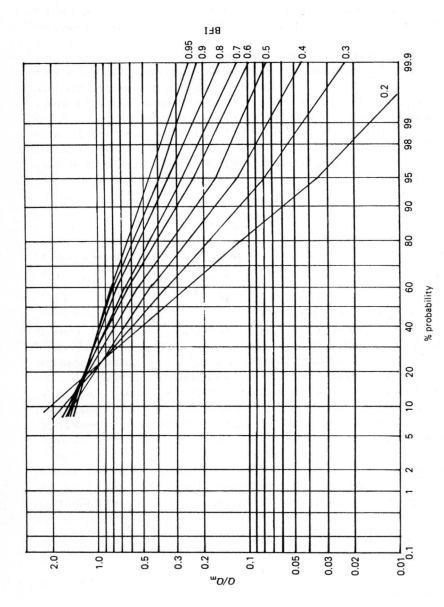

Figure 6.24   *Typical flow duration curves for values of BFI (by Dr A. Gustard, after Low Flow Studies* [26]

If the catchment is ungauged then Q_m must be derived by the use of rainfall records and estimates of evaporation. In the long term, catchment runoff may be assumed to be the difference between annual average rainfall ($SAAR$) and actual evaporation (AE), since infiltration and percolation will appear eventually as baseflow.

In the UK, $SAAR$ is estimated from the 1:625,000 annual average rainfall map for 1941-70 and potential evaporation (PE) from the 1:2 million map of annual average PE based on the Penman equation, both available from the Meteorological Office. Elsewhere PE may be calculated from weather station data as described in chapter 3.

Actual evaporation (AE) then has to be estimated from potential evaporation by using an adjustment factor ϕ (table 6.1). This factor is based on actual data of rainfall and runoff for stations in the UK with more than 10 years of data, and refers to the UK. In other parts of the world it may need modification, using similar methods.

TABLE 6.1 *Adjustment factor for estimating actual evaporation* [2]

$$AE = \phi \times PE$$

$SAAR$	500	600	700	800	900	1000	1100
ϕ (mm)	0.88	0.90	0.92	0.94	0.96	0.98	1.0

When $SAAR$ exceeds 1100 mm, AE is assumed to equal PE, which is related to a freely transpiring grass surface. Periods when AE is limited by soil moisture deficit are assumed to be compensated for by AE exceeding PE at other times by evapotranspiration from other types of vegetation exceeding that of grass.

If local data are available from a similar neighbouring gauged catchment the runoff rate may be used directly, but if $SAAR$ for the gauged catchment is different from the one under study, its loss rate in mm should be deducted from the $SAAR$ for the latter.

References

1. YOUNG, L. H. Mean annual rainfall/run-off relationship. *J. Inst. Wat. Eng.*, **24**, No. 7 (1970) 423
2. Natural Environment Research Council. *Flood Studies Report*. Vol. 1, NERC, 1975, Chapter 6
3. B.S. 3680 Methods of measurement of liquid flow in open channels. Part 3: 1964 Velocity area methods. Part 4: 1965 Weirs and flumes
4. CORBETT, D. M. *et al.*, Stream-gaging procedure. *Water Supply Paper 888*, Washington D.C., 1945
5. BUCHANAN, T. J. and SOMERS, W. P. Discharge measurements of gaging stations. *U.S. Geological Survey Tech. Water Resources Inv.*, Bk. 3, 1969
6. CARTER, R. W. and DAVIDIAN, J. Discharge ratings at gauging stations. *U.S. Geological Survey Surface Water Tech.*, Bk. 1, 1965
7. U.S. Bureau of Reclamation. *Water Measurement Manual*, 1953

8. ACKERS, P. and HARRISON, A. J. M. Critical depth flumes for flow measurement in open channels. *Hydrological Research Paper No. 5*, H.M.S.O., London, 1963
9. PARSHALL, R. L. Measuring water in irrigation channels with Parshall flumes and small weirs. *U.S. Dept. Agr. Circular 843*, 1950
10. ACKERS, P. Flow measurement by weirs and flumes. *Int. Conf. on Modern Developments in Flow Measurement*, Harwell 1971, Paper No. 3
11. WHITE, W. R. Flat-vee weirs in alluvial channels. *Proc. Am. Soc. Civ. Eng.*, 97, HY3 (March 1971) 395–408
12. WHITE, W. R. The performance of two dimensional and flat-V triangular profile weirs. *Proc. Inst. Civ. Eng.*, Suppl. (ii), (1971) 21–48
13. BURGESS, J. S. and WHITE, W. R. Triangular profile (Crump) weir: two dimensional study of discharge characteristics. *Report No. INT 52, Institute of Hydrology*, Wallingford, United Kingdome, 1952
14. HARRISON, A. J. M. and OWEN, M. W. A new type of structure for flow measurement in steep streams, *Proc. Inst. Civ. Eng.*, (1967) 273–96
15. SMITH, C. D. Open channel water measurement with the broad-crested weir. *Int. Commun. Irr. Drainage Bull.*, (1958) 46–51
16. HOSEGOOD, P. H. and BRIDLE, M. K. A feasibility study and development programme for continuous dilution gauging. *Report No. 6, Institute of Hydrology*, Wallingford, United Kingdom
17. ISO/R 55, 1966 Liquid flow measurement in open channels; dilution methods for measurement of steady flow. Part 1, Constant rate injection
18. LITTLEWOOD, I. G. *Research and development of a streamflow dilution gauging technique for the Llyn Brianne Acid Waters Study, Wales*, Department of Geography, University College, Swansea, 1986
19. AASTAD, J. and SØGNEN, R. Discharge measurements by means of a salt solution; the relative dilution method. *Proc. IASH General Assembly, Rome 1954*, Vol. 3, pp. 289–292
20. HERSCHY, R. W. and LOOSEMORE, W. R. The ultrasonic method of river flow measurement. *Symp. on River Gauging by Ultrasonic and Electromagnetic Methods*, University of Reading, Dec. 1974
21. FOSTER, W. E. Experience with the construction and engineering operation of ultrasonic gauging stations. *Symp. on River Gauging by Ultrasonic and Electromagnetic Methods*, University of Reading, Dec. 1974
22. BOYER, M. C. Determining discharge of gauging stations affected by variable slope. *Civ. Eng.*, 9, (1939) 556
23. MITCHELL, W. D. Stage–fall–discharge relations for steady flow in prismatic channels. *U.S. Geological Survey Water Supply Paper 1164*, Washington D.C., 1954
24. STEVENS, J. C. A method of estimating stream discharge from a limited number of gaugings. *Eng. News*, 18 July 1907
25. KOELZER, V. A. Reservoir Hydraulics. *Handbook of Applied Hydraulics* (ed. by C. V. Davis and K. E. Sorenson), 3rd editon, McGraw-Hill, New York, 1969, Section 4
26. *Low Flow Studies*, Institute of Hydrology, Wallingford, United Kingdom, January 1980
27. University of Salford, Department of Civil Engineering. *Small-scale hydroelectric potential of Wales*, Department of Energy, London, 1980

Further reading

HERSCHY, R. W. New methods of river gauging. *Facets of Hydrology* (ed. J. C. Rodda), John Wiley, New York, 1976, Chapter 5

HORTON, R. E. Erosional development of streams and their drainage basins. *Bull Geol. Soc. Am.*, **56** (March 1954) 275

Logarithmic plotting of stage–discharge observations. *Tech. Note 3. Water Resources Board*, Reading, 1966.

NASH, J. E. and SHAW, B. L. Flood frequency as a function of catchment characteristics. *Inst. Civ. Eng. Symposium on River Flood Hydrology*, 1966, pp. 115–6

RODDA, J. C. The significance of characteristics of basin rainfall and morphometry in a study of floods in the United Kingdom. *Int. Assoc. Sci. Hydrol. Symposium on Floods and their Computation, Leningrad. International Association of Scientific Hydrology*, **85** (1967) 834

SMOOT, G. F. and NOVAK, C. E. Measurement of discharge by the moving boat method. *U.S. Geological Survey Tech. Water Resources Inv.*, Bk. 3 (1969) A11

STRAHLER, A. N. Statistical analysis of geomorphic research. *J. Geol.*, **62**, No. 1 (1954)

Problems

6.1   A river gauging gives $Q = 4010$ m^3/s. The gauging took 3 h during which the gauge fell 0.15 m. The slope of the river surface at the gauging site at the time was 80 mm in 500 m, and the cross-section approximated a shallow rectangle 200 m wide by 11 m deep. What adjusted value of discharge would you use? What value of n in Manning's formula results?

6.2   The following discharge observations have been made on a river

Gauge height (ft.)	Measured discharge (ft.^{3}/s × 1000)	Rise + or fall − (ft./h)
10.4	50	−
12.2	65	−
13.9	77	−
14.3	80	−
22.3	150	−
27.3	180	−0.32
28.1	228	+0.80
30.8	256	+0.525
32.6	225	−0.36
35.2	251	−0.355
38.9	338	+0.345
40.3	316	−0.22
40.8	352	+0.18
41.5	333	−0.235
42.2	362	−

Using Boyer's method adjust the figure for slope variation to produce a steady flow discharge rating curve for the river.

6.3 Explain how observations of river discharge at particular gauge heights can be corrected so that they fall on a smooth curve, and explain why this is desirable.

A river discharge was measured at $Q = 2640 \, \text{m}^3/\text{s}$. During the 100 minutes of the measurement the gauge height rose from 50.40 to 50.52 m. Level readings upstream and downstream differed by 100 mm in 700 m. The flood wave celerity was 2.2 m/s. Give the corrected rating curve co-ordinates.

6.4 An unregulated stream provides the following volumes over an 80-day period at a possible reservoir site

Day	Runoff volume $(m^3 \times 10^6)$	Day	Runoff volume $(m^3 \times 10^6)$	Day	Runoff volume $(m^3 \times 10^6)$
0	0	28	0.7	56	0.6
2	2.0	30	0.8	58	1.2
4	3.2	32	0.8	60	1.4
6	2.3	34	0.7	62	1.8
8	2.1	36	0.7	64	2.0
10	1.8	38	0.5	66	2.3
12	2.2	40	0.4	68	3.2
14	0.9	42	0.7	70	3.4
16	0.5	44	0.8	72	3.5
18	0.3	46	0.4	74	3.7
20	0.7	48	0.3	76	2.8
22	0.7	50	0.2	78	2.4
24	0.6	52	0.2	80	2.0
26	1.2	54	0.4		

(a) Plot the data in the form of a mass diagram.
(b) Determine average, maximum and minimum flow rates.
(c) What reservoir capacity would be needed to ensure maintenance of average flow for these 80 days if the reservoir is full to start with?
(d) How much water would be wasted in spillage in this case?

6.5 The average domestic per capita demand for water in an expanding community is $0.20 \, \text{m}^3/\text{day}$. Industrial demand is 30 per cent of total domestic requirements. The town has 100 000 inhabitants now and is expected to double its population in future.

Water is supplied from a river system with existing storage capacity of $10^7 \, \text{m}^3$ and whose mean daily discharges for each month of the year are as follows (thousands of m^3)

Jan.	Feb.	March	April	May	June	July	Aug.	Sept.	Oct.	Nov.	Dec.
290	250	388	150	64.5	50	64.5	117	283	388	317	385

Compensation water of 1.5 m^3/s is to be provided constantly.

Find, to a first approximation and for an average year, the additional storage capacity that will have to be provided if the population doubles in size. Also determine the quantity of water spilled to waste in such a year and compare it with the wastage now. Assume the existing storage is half full on 1 January.

6.6   A community of 60 000 people is increasing in size at a rate of 10 per cent per annum. Average demand per head (for all purposes) is currently 0.20 m^3/day and rising at a rate of 5 per cent per annum. The existing water supply has a safe yield of 0.5 m^3/s. A river is to be used as an additional source of supply. Its mean daily discharges for each month of the water-year are listed below in thousands of m^3

April	May	June	July	Aug.	Sept.	Oct.	Nov.	Dec.	Jan.	Feb.	March
220	250	370	670	865	1630	670	530	270	300	280	280

Allowing for compensation water of 3 m^3/s from October–March inclusive and 5 m^3/s from April–September inclusive, determine to a first approximation the storage capacity required on the river to ensure the community's water supply 20 years from now, assuming present trends continue, and that the reservoir would be full at the end of November.

6.7   List eight characteristics of drainage basins affecting their discharge hydrographs and comment on each.

6.8   A river gauging gives Q = 2060 m^3/s. The gauging took 2 h during which the gauge fell 0.12 m. The river surface slope at the time was 5 cm in 400 m and the cross-section of the river at the site was 300 m wide by 4 m deep.

What adjusted discharge would you use? What is Manning's n for this river at this time, and what does the value indicate about the river's condition?

6.9   The two-year monthly discharges of a river into a reservoir are listed below

Month	Monthly discharge ($m^3 \times 10^3$)	
	Year 1	Year 2
January	576	102
February	658	308
March	287	432
April	329	533

May	370	390
June	247	287
July	102	164
August	21	123
September	21	123
October	21	141
November	41	183
December	83	221

Draw a mass diagram of inflow and determine the following.

(a) If the reservoir is full at the end of February, year 1, what is the permissible draw-off in l/s so that the reservoir may be full at the end of June, year 2?

(b) If the reservoir is full at the beginning of January, year 1, and the draw off is 60 l/s for the first year and 80 l/s for the second year, what will be the state of storage in the reservoir at the end of December, year 2?

(c) What storage capacity will be required in the reservoir in the second case to ensure the required discharges may be supplied?

6.10 An unregulated river has monthly mean flows (in m^3/s) as follows:

Jan.	Feb.	March	April	May	June	July	Aug.	Sept.	Oct.	Nov.	Dec.
5.4	8.3	9.1	8.8	6.3	6.9	10.2	13.7	19.4	16.7	11.0	21.9

Allowing compensation water of 4.0 m^3/s and reservoir losses of 0.5 m^3/s, what storage capacity of reservoir is required to ensure that, on average, no water is spilled? What would the average net yield of the reservoir then be? Assume 30-day months.

6.11 (a) Draw a flow duration curve from the tabulated data below —the mean monthly discharge (flows in m^3/s) of a source of water.

	Year 1	Year 2	Year 3
January	110	180	193
February	102	118	109
March	97	88	99
April	84	79	91
May	70	56	82
June	62	52	74
July	45	47	68
August	67	35	43

September	82	60	30
October	134	75	48
November	205	98	49
December	142	127	63

If a hydro-power plant is to be developed at the site to which the data refers and where the head available is 15 m, what would be a reasonable first estimate of plant capacity and annual energy production on the basis of the data? Justify your choice.

(b) What is the mean monthly flow with a return period of once in 10 years?

6.12 The monthly inflow (in millions of m^3) to a reservoir with 100 km^2 surface area is listed below for a 24 month period.

Month	Year 1	Year 2	Month	Year 1	Year 2
1	35	38	7	14	6
2	28	30	8	17	15
3	25	24	9	23	20
4	16	12	10	27	28
5	10	8	11	36	40
6	9	7	12	40	42

Losses from evaporation are assumed to be 0.10 m/month. Compensation water of 0.3 m^3/s is constantly released.

If the reservoir was full at the end of month 3, year 1, and full again at the end of month 12, year 2, calculate

(a) the constant net yield over the period of 2 years
(b) the total water spilled
(c) the storage capacity necessary.

6.13 It is proposed to develop a waterfall for hydro-electric power. The flow duration curve is given in tabular form below. The available head is 70 m.

Make a first estimate of installed capacity and annual energy production.

% time equalled or exceeded	10	20	30	40	50	60	70	80	90	100
Q (m^3/s)		4.5	3.5	2.9	2.5	2.2	1.9	1.6	1.4	1.1	0.5

7 Hydrograph Analysis

7.1 Components of a natural hydrograph

The various contributing components of a natural hydrograph are shown in figure 7.1. To begin with there is *baseflow* only; that is, the groundwater contribution from the aquifers bordering the river, which go on discharging more and more slowly with time. The hydrograph of baseflow is near to an exponential curve and the quantity at any time is represented very nearly by

$$Q_t = Q_0 e^{-\alpha t}$$

where Q_0 = discharge at start of period
Q_t = discharge at end of time t
α = coefficient of aquifer
e = base of natural logarithms.

As soon as rainfall begins there is an initial period of interception and infiltration before any measurable runoff reaches the stream channels, and during the period of rain these losses continue in a reduced form as discussed previously; so the *raingraph* has to be adjusted to show *net*, or *effective rain*. When the initial losses are met, surface runoff begins and continues to a peak value, which occurs at a time t_p, measured from the centre of gravity of the raingraph of net rain. Thereafter it declines along the *recession limb* until it completely disappears. Meantime the infiltration and percolation that has been continuing during the gross rain period results in an elevated groundwater table, which therefore contributes more at the end of the storm flow than at the beginning but thereafter is again declining along its *depletion curve*.

Surface runoff is, for convenience, assumed to contain two other components: *channel precipitation* and *interflow*. Channel precipitation is that portion of the total catchment precipitation that falls directly on the stream, river and lake surfaces. It is usually small but if large lakes are present in the catchment it can be quite important and then requires separate treatment. Interflow refers to water travelling horizontally through the upper horizons of the soil, perhaps in

150

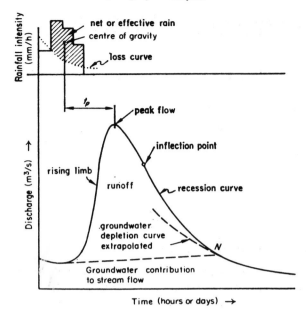

Figure 7.1 *Component parts of a natural hydrograph*

artificial tile drain systems or above *hard-pans* or impermeable layers immediately below the surface. Such flow can vary from nothing to appreciable fractions of total runoff.

Since the groundwater contribution to flood flow is quite different in character from surface runoff it should be analysed separately, and one of the first requirements in hydrograph analysis therefore is to separate these two.

7.2 The contribution of baseflow to stream discharge

Since baseflow represents the discharge of aquifers, changes occur slowly and there is a lag between cause and effect that can easily extend to periods of days or weeks. This will depend on the *transmissibility* of the aquifers bordering the stream and the climate. Some of the infinite number of natural conditions are considered below.

A broad distinction should be made between *influent* and *effluent* streams. An influent stream is one where the baseflow is negative; that is the stream feeds the groundwater instead of receiving from it (for example, irrigation channels operate as influent streams and many natural rivers that cross desert areas also do so). The negative contribution is taking place at the expense of contributing aquifers on other parts of the stream, since there can be no baseflow from a *wholly* influent stream. Such a stream (for example, a Middle Eastern *wadi*) will dry up completely in rainless periods and is called *ephemeral*; it has a hydrograph of the form of figure 7.2.

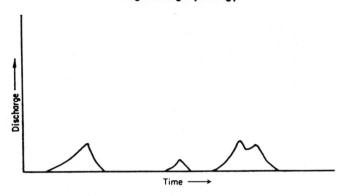

Figure 7.2 *Hydrograph of an ephemeral stream*

An effluent stream on the other hand is fed by the groundwater and acts as a drain for bordering aquifers. The great majority of streams in Britain and Europe are in this category.

Intermittent streams are those that act as both influent and effluent streams according to season, tending to dry up in the dry season.

Perennial streams are greatly in the majority, with a low dry-season flow fed by baseflow, and are mainly effluent streams, though many perennial rivers crossing different geological formations of varying permeability and subject to different climates are both influent and effluent at different parts of their courses. A good example of this is the river Euphrates in Iraq. Figure 6.18 shows a part-annual hydrograph of the river Euphrates and the slow seasonal variation of the baseflow can be observed. This baseflow is derived principally from the headwaters of the catchment in northern Iraq, Turkey and Syria. At Hit, where the hydrograph was observed, the river for much of the year is influent.

Bank storage describes the portion of runoff in a rising flood that is absorbed by the permeable boundaries of a water course above the normal phreatic surface, it is illustrated in figures 7.3 and 7.4. In the latter figure the direction of the arrows showing influx of groundwater to the stream will be reversed during the flood period while the surface level of the stream is above the phreatic

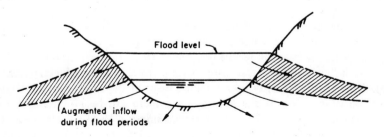

Figure 7.3 *Influent stream*

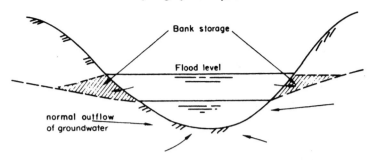

Figure 7.4 *Effluent stream*

surface. As a result the hydrograph of a particular flood might well have a base-flow contribution as indicated in figure 7.5. Such a separation as is shown there is virtually impossible to make quantitatively but it is qualitatively correct.

In many natural rivers, depending naturally on bank permeability and the slope of the phreatic surface, the variation in baseflow will be much less than indicated in figure 7.5 and will cause only a slight dip from the extrapolation of the depletion curve, followed by a gradual rise to a higher-than-initial value as indicated in figure 7.6.

7.3 Separation of baseflow and runoff

It has been shown in the preceding paragraphs that the dividing line between runoff and baseflow is indeterminate and can vary widely. Since, to *analyse* its

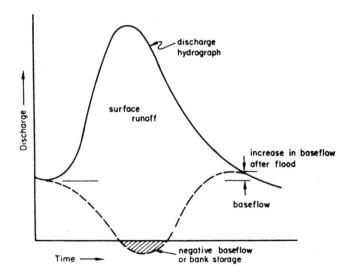

Figure 7.5 *Negative baseflow*

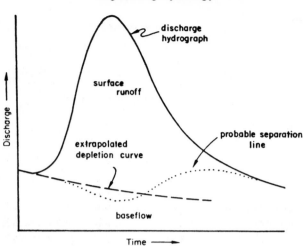

Figure 7.6   *Baseflow separation*

precise position would require a detailed knowledge of the geohydrology of the catchment, including the areal extent and transmissibility of the aquifers, it is generally more practical to use a consistent separation technique. Which of the following is used depends on the data available.

If a continuous discharge record of the stream, over a period of a few years, is available, the hydrograph should be plotted in the manner of figure 7.7(a). This is examined for portions that include recession curves running into baseflow contribution only, after runoff has ceased, at as many different stages as possible. These sections are abstracted from the continuous hydrograph and plotted again to a log Q vertical scale and linear time scale, as shown in figure 7.7(b). Starting with the lowest recession flow in the record, a curve is now constructed that is tangential to the lower portions (that is, the true depletion curves) of the log Q abstracted plots. This is most easily done by moving tracing paper over the plots, with the abscissae coincident, until each log Q plot in successive increasing magnitude fits into the growing curve and extends it fractionally upward. The tangential curve thus established to the highest stage possible is then converted back to the linear vertical scale and is called the *master depletion curve* for the particular gauging station. It can now be applied to the hydrograph of a particular storm period in the manner indicated in figure 7.8 whereby the depletion curves are fitted together at their lower ends and the point of divergence marked as N. N represents the point at which surface runoff has effectively ceased and a straight line is drawn to it from the point of sudden rise. This line, shown dashed in figure 7.8, represents the base line of the hydrograph of surface run-off, which can then be analysed.

While the procedure outlined above is probably the best available, it does depend on previously observed data that are not always available. An alternative

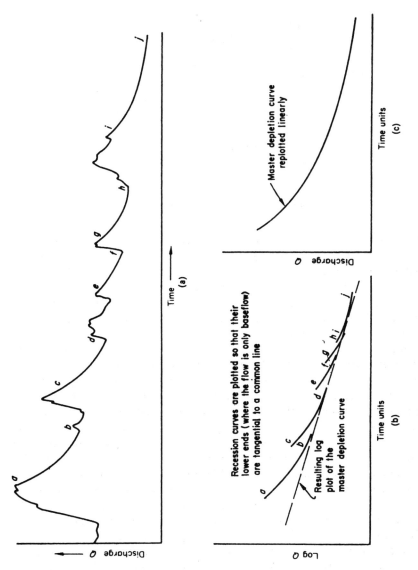

Figure 7.7 Derivation of a master depletion curve. (a) Normal hydrograph with recession curves selected. (b) Log plot of recession curves. (c) Linear plot of master depletion curve

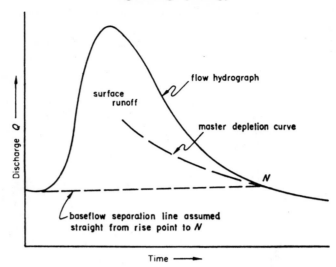

Figure 7.8 *Procedure to separate baseflow*

procedure is to establish the point of greatest curvature on the recession limb of the hydrograph. This is perhaps most easily done by computing the ratio between Q at any time and say 2 h (or any convenient interval) later. An example will illustrate the method. Figure 7.9 is the observed hydrograph of a river over a period of several days. It is desired to separate surface runoff and baseflow.

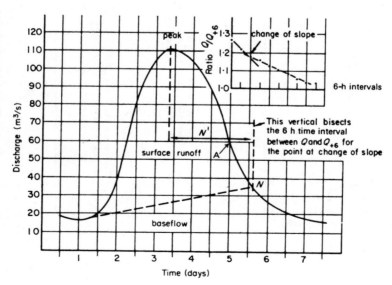

Figure 7.9 *Alternative method of separating baseflow*

Starting at point A and using a 6-h separation for successive ratios, the computations are as shown in table 7.1.

It can be seen from the inset graph of the ratio Q/Q_{+6} against time interval on figure 7.9 that two separate slopes are apparent, the upper being associated with runoff and the other with groundwater depletion. At their intersection, the

TABLE 7.1 *Computation to assist in finding N*

Day	Hour	Q (m^3/s)	Q_{+6} (m^3/s)	Ratio Q/Q_{+6}
5	1200	60.1	47.5	1.27
	1800	47.5	39.0	1.22
	2400	39.0	33.2	1.18
6	0600	33.2	28.6	1.16
	1200	28.6	25.2	1.13
	1800	25.2	22.7	1.11
	2400	22.7	20.9	1.09
7	0600	20.9	19.7	1.06
	1200	19.7	18.9	1.04
	1800	18.9	18.2	1.04
	2400	18.2	—	

critical ratio can be determined and the first point beyond the region of intersection on the groundwater side gives a conservative position for N. The subsequent hydrograph analysis is not very sensitive to the precise position of N and either 0300 h or 0600 h on day 6 would be satisfactory. A straight line is now drawn to N from the point where the hydrograph started to rise, as before, The total amount of runoff can now be obtained by measuring the area under the curve and above the straight line.

The position of N is important also in synthesising hydrographs, as will be seen in section 7.11, since it partly defines the *baselength* of the hydrograph. The baselength is made up of the part before the peak, which depends on the length of the period of rain and t_c the period of concentration, and the recession limb after the peak, which depends primarily on the character of the catchment. From observations on many natural catchments the position of N can be established empirically from table 7.2 in days after the peak of the flood.

7.4 Evaluation of baseflow. A method of evaluating baseflow for an ungauged catchment is given in reference 1. Through regression analysis of CWI (see section 4.4.4) and catchment characteristics of a large number of British catchments, the following equation is proposed for the 'average non-separated flow' or ANSF

$$\text{ANSF} = (3.26 \times 10^{-4})\,(\text{CWI} - 125) + (7.4 \times 10^{-4})\,\text{RSMD} + (3 \times 10^{-3})$$

$$(7.1)$$

TABLE 7.2   *Catchment area as a guide to N*

Catchment area (km^2)	Time from peak to N (days)
250	2
1250	3
5000	4
12500	5
25000	6

where ANSF is baseflow in m^3/s/km^2 and RSMD is an index of flood potential dependent on climate, and is defined in section 9.4.

7.5 The unit hydrograph

Having derived the hydrograph of surface runoff by the methods discussed in preceding sections, the problem now arises of how it can be correlated with the rainfall that caused it. Clearly the quantity and intensity of the rain both have a direct effect on the hydrograph but it has not yet been made clear how, and to what extent, each of these affects it. The method of doing this is a part-empirical technique that uses the concept of the *unit hydrograph* (also called the *unit-graph*), first described by Sherman [2].

It should be emphasised that the correlation sought is between the *net* or *effective rain* (that is, the rain remaining as runoff after all losses by evaporation, interception and infiltration have been allowed for) and the surface runoff (that is, the hydrograph of runoff minus baseflow).

The method involves three principles, which are as follows.

1. With uniform-intensity net rainfall on a particular catchment, different intensities of rain of the same duration produce runoff for the same period of time, although of different quantities. This is an empirical rule that is approximately true and is illustrated in figure 7.10.
2. With uniform-intensity net rain on a particular catchment, different intensities of rain of the same duration produce hydrographs of runoff, the ordinates of which, at any given time, are in the same proportion to each other as the rainfall intensities. That is to say, that n times as much rain in a given time will give a hydrograph with ordinates n times as large. In figure 7.10 the ordinates at time t_1 are np and p respectively for rainfall intensities of ni and i.
3. The principle of superposition applies to hydrographs resulting from contiguous and/or isolated periods of uniform-intensity net rain. This is illustrated in figure 7.11 where it can be seen that the total hydrograph of runoff due to the three separate storms is the sum of three separate hydrographs.

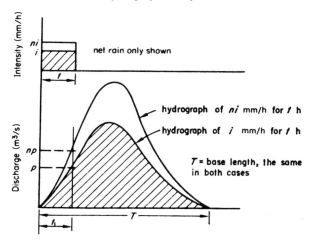

Figure 7.10 *Proportional principle of the unitgraph*

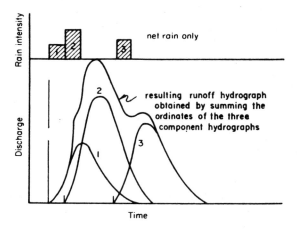

Figure 7.11 *Principle of superposition applied to unitgraphs*

Having established these principles the concept of unit rain is now introduced. A unit of rain can be any specified amount, measured as depth on the catchment, usually 1 cm or 1 in. but not exclusively so. The unit rain then must all appear as runoff in the *unit hydrograph*. The area under the curve of the hydrograph has the dimensions of instantaneous discharge multiplied by time, or

$$\frac{L^3}{T} \times T = L^3 = \text{volume of runoff}$$

so that although unit rain is spoken of as 1 cm over the whole of the catchment area the resulting runoff is given in cubic metres, and the quantities involved are identical. If the unitgraph for a particular catchment, *and a particular duration of rain* is known, then from principle 2, the runoff from any other rain of the same duration can be predicted.

This is a first step towards the complete correlation sought, but if the rainfall should be of different duration from that of the unitgraph then the unitgraph must be altered before it can be used.

7.6 Unit hydrographs of various durations

7.6.1 Changing a short duration unitgraph to a longer duration unitgraph. The simplest way to produce a unitgraph for a longer duration of rain is illustrated in figure 7.12.

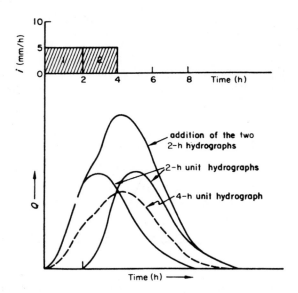

Figure 7.12   *Changing a short period unitgraph to a long period one (if the long is an even multiple of the short)*

Suppose a 2-h unitgraph is given and a 4-h unitgraph is wanted. This can be obtained by assuming a further 2-h period of net rain immediately following the first, which will give rise to an identical unitgraph but shifted to the right in time by 2 h. If the two 2-h unitgraphs are now added graphically, the total hydrograph obtained represents the runoff from 4 h of rain at an intensity of $\frac{1}{4}$ cm/h. (This must be so because the 2-h unitgraph contains 1 cm rain.) This total hydrograph is therefore the result of rain at twice the intensity required and so the 4-h unitgraph is derived by dividing its ordinates by 2. This is shown as the dashed line on figure 7.12. It will be observed that it has a longer time base by 2 h than the 2-h unitgraph; this is reasonable since the rain has fallen at a lower intensity for a longer time.

7.6.2 Changing a long duration unitgraph to a shorter duration unitgraph. To derive a short-rain period unitgraph from that for a longer period it is necessary to use an S-curve technique. An S-curve is simply the total hydrograph resulting from a series of continuous uniform-intensity storms delivering 1 cm in t_1 h on the catchment; that is, it is the hydrograph of runoff of continuous rainfall at an intensity of $1/t_1$. Such a hydrograph has the form of figure 7.13, the discharge of the catchment becoming constant after t_c, the time of concentration, when every part of the catchment is contributing and conditions are in a steady state. Thus each S-curve is unique for a particular unitgraph duration, in a particular drainage basin.

If a second S-curve is drawn one unit period to the right of the first, then clearly the difference between the two S-curves expressed graphically equals the runoff of one t_1 h unitgraph.

If the unitgraph for a short period storm of t_2 h is required, it can be obtained by drawing the S-curve again, but shifted only t_2 h along the time axis. The graphical difference between ordinates of the two S-curves now represents the runoff of t_2 h rain at an intensity of $1/t_1$ cm/h. The ordinates of this S-curve difference graph must therefore be multiplied by t_1/t_2 so that the rain intensity represented is $1/t_2$ cm/h, which is the intensity required for the t_2 unitgraph. The procedure is illustrated in figure 7.13.

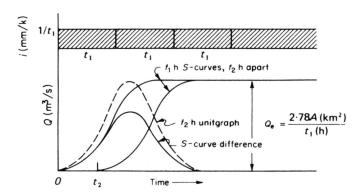

Figure 7.13 *Transposing unitgraphs by S-curves*

If the time base of the unitgraph is T h, then steady-state runoff must occur at T h and so only T/t_1 unitgraphs are necessary to develop constant outflow and so produce an S-curve. The equilibrium flow, Q_e, can easily be obtained since 1 cm on the catchment is being supplied and removed every t_1 h:

$$Q_e = \frac{2.78A}{t_1} \qquad \text{or} \qquad Q_e = \frac{645A}{t_1}$$

where A is catchment area (km^2)   where A is catchment area (mile^2)

        t_1 is duration (h) t_1 is duration (h)

and Q_e is in m^3/s.   and Q_e is in ft.^{3}/s.

It will be apparent that the method can be used for altering the unit period either way, longer or shorter, and that if changing from shorter to longer duration then t_2 need not be a direct multiple of t_1. Although the method has been described graphically, in practice its application is usually made in tabular form and example 7.1 illustrates it.

Example 7.1. Given the 4-h unit hydrograph listed in column (2) of table 7.3 derive the 3-h unit hydrograph. The catchment area is 300 km^2.

TABLE 7.3   *S-curve method (all values except those in columns (1) are in units of m^3/s)*

(1)	(2)	(3)	(4)	(5)	(6)	(7)
Time (h)	4-h unitgraph	S-curve additions	S-curve columns (2) + (3)	lagged S-curve	Column (4) minus column (5)	Column 6 × (4/3) = 3-h unitgraph
0	0	—	0	—	0	0
1	6	—	6	—	6	8
2	36	—	36	—	36	48
3	66	—	66	0	66	88
4	91	0	91	6	85	113
5	106	6	112	36	76	101
6	93	36	129	66	63	84
7	79	66	145	91	54	72
8	68	91	159	112	47	63
9	58	112	170	129	41	55
10	49	129	178	145	33	44
11	41	145	186	159	27	36
12	34	159	193	170	23	31
13	27	170	197	178	19	25
14	23	178	201	186	15	20
15	17	186	203	193	10	13.5[a]
16	13	193	206	197	9	12[a]
17	9	197	206	201	5	6.5[a]
18	6	201	207	203	4	5.5[a]
19	3	203	206	206	0	0[a]
20	1.5	206	207	206	1	1.5[a]
21	0	206	206	207	− 1	

[a]Slight adjustment is required to the tail of the 3-h unitgraph. This is most easily done by eye (see figure 7.14).

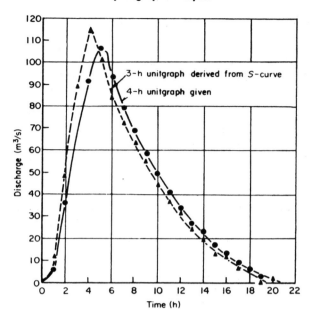

Figure 7.14 *Unitgraph derived by S-curve method*

The S-curve equilibrium flow $Q_e = (2.78 \times 300)/4 = 208 \ m^3/s$.

It will be noted that $Q_e = 208 \ m^3/s$, as calculated, agrees very well with the tabulated S-curve terminal value 207. This is an indication that the 4-h period of the unitgraph is correctly assessed. Very often with an uneven rainfall distribution, an attempt has to be made to reduce the net rain to a uniform-intensity rain of a particular duration. The S-curve can in this way serve as a check on the chosen value. If the S-curve terminal value had fluctuated wildly and not steadied to a minor variation it would have indicated an incorrect rainfall-time for the unitgraph.

Note also that it was not necessary in table 7.3 to set out T/t_1 columns of the 4-h unitgraphs, and add them laterally. The *S-curve additions* are the S-curve ordinates shifted in time by 4 h. Since the first 4 h of unitgraph and S-curve are the same, the S-curve additions and S-curve columns are filled in, in alternate steps. The effect is the same as setting out rows of unitgraph ordinates successively staggered by 4 h, since the S-curve additions represent the sum of all previous unitgraph ordinates.

7.7 The unit hydrograph as a percentage distribution

The distribution graph, first used by Bernard [3], represents the unitgraph in the form of percentages of total flow occurring in particular unit periods. Since the discharge represented by a unitgraph is directly proportional to net rain, the percentages in unit times will remain constant whatever the net rain. This is a useful means of applying the unitgraph method in some cases.

In figure 7.15 a unit hydrograph is shown, together with the derived distribution graph that represents it. The areas under the curve and under the step

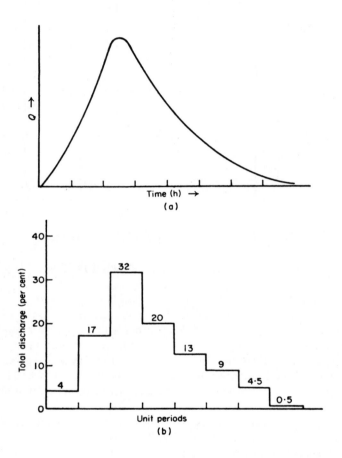

Figure 7.15   (a) *Unit hydrograph.* (b) *Derived distribution graph*

line are the same and so, in deriving unitgraphs from distribution percentages, a smooth line must be drawn through the steps to give equal areas.

The distribution graph is therefore less precise than the hydrograph but is much better suited for iterative processes of derivation, as will be seen in section 7.8.

7.8 Derivation of the unit hydrograph

The unitgraph for a particular catchment can be derived from the natural hydrograph resulting from any storm that covers the catchment and is of reasonably uniform intensity. If the catchment is very large that is, greater than (say) 5000 `km²`), it may never be covered by a uniform-intensity storm, since these are limited in size by meteorological conditions. In such a case the catchment should be divided up into tributary catchments and the unitgraphs for each of these determined separately.

The first step is to separate the baseflow from surface runoff (section 7.3) and plot the runoff and the rain graph on the same time base. The quantity of net storm rain must then be estimated and its intensity and duration established. A check is now made on the quantity of net rain on the catchment and the amount of runoff under the hydrograph. These should be the same and one or the other may require adjustment.

The unitgraph can now be obtained by dividing the runoff hydrograph ordinates by the net rain in cm. The adjusted ordinates represent the unitgraph for the particular duration established.

It is always advisable to determine several unitgraphs, using separate and distinct isolated uniform-intensity storms, if available. Natural events like rainstorms and runoff are affected by a multiplicity of factors and no two are precisely the same. Frequently the best natural data will be for different rain durations and the resulting unitgraphs will require to be altered to the same duration (section 7.6). Once a number of such hydrographs has been obtained for the same duration, an 'average' or typical unitgraph can be determined as shown in figure 7.16. The ordinates are *not* averaged since this would produce an untypical peak. The peak values of the separate unitgraphs are averaged as are the values of the time from the beginning of runoff to the peak. These values are assigned to the average unitgraph which is then sketched in to a median form on both rising and falling limbs, so that the total area under the curve is equal to 1 cm runoff.

7.9 Unit hydrographs from complex or multi-period storms

While the approach outlined in section 7.8 is simple and direct, it presupposes that the records contain a number of isolated, uniform-intensity storms and the corresponding natural hydrographs. Frequently this is not the case and methods are required for deriving unitgraphs from more complex storms, involving varying intensities of rain with runoff hydrographs consisting of several superimposed separate storm hydrographs.

To derive unitgraphs from such records is more laborious than for simple storms but a variety of methods is available, two of which are discussed below.

The first, described by Linsley *et al.* [4] requires the writing and successive solving of a series of equations for each ordinate of the complex hydrograph,

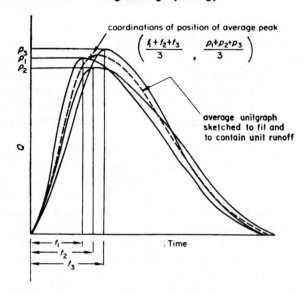

Figure 7.16 *Average unit hydrograph from a number of derivations for one catchment*

baseflow being assumed previously separated. The process may be illustrated by reference to figure 7.17.

The first rain period, of duration and intensity t and i_1 respectively, gives rise to runoff illustrated by the hypothetical hydrograph bounded by the lower dashed line. Each ordinate of this hydrograph is ti_1 times the unitgraph ordinate $U_1, U_2 \ldots U_n$. Similarly, the second and third rains of intensity i_2 and i_3 respectively produce additional runoff whose ordinates in each case are ti_2 and ti_3, multiples of the t h unitgraph shifted appropriately in time. If the complex hydrograph is now defined by ordinates at suitable equal intervals (conveniently but not essentially fixed as a whole multiple of t h) then the first ordinate of the unitgraph, U_1, is obtained from $Q_1 = ti_1U_1$ where Q_1 is the observed runoff, hence U_1 can be found. For the second ordinate, $Q_2 = ti_1U_2 + ti_2U_1$ in which equation U_2 is the only unknown.

The third ordinate is similarly obtained from $Q_3 = ti_1U_3 + ti_2U_2 + ti_3U_1$ where U_3 is now the only unknown. Proceeding in this way, the t h unitgraph ordinates can be successively determined.

In the above illustration all the rain periods, although of different intensities, were assumed to be of the same duration t h. This is a condition for the use of the method since otherwise other variables U'_1, U'_2 etc. (the ordinates of the t' h unitgraph) would be introduced.

Although the method appears simple, since each U ordinate depends on the preceding ones, and on the assumptions about intensity and duration of rainfall

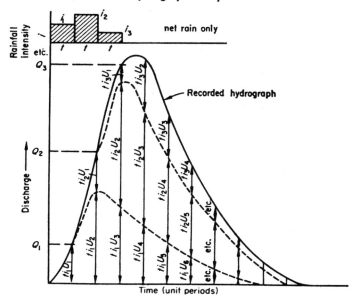

Figure 7.17   *Derivation of unit hydrograph from a multi-period storm hydrograph*

and deduction of an assumed baseflow, errors accumulate and several trials and restarts may be necessary to find a reasonable unitgraph.

The second method is due to Collins [5] and is amongst the simplest of various iterative methods proposed. To illustrate its use a unit hydrograph will be derived from the data of rainfall and natural discharge for the catchment of the River Rother at Woodhouse Mill in Yorkshire.

Collins' method for the determination of a unit hydrograph from a multi-period storm. This method requires the initial selection of a set of coefficients, or percentage distributions (see section 7.7) of the unitgraph. This distribution graph is then applied to the various rain periods, excepting the largest one, and the resulting discharge subtracted from the actual discharge to obtain a set of 'residuals'. These residuals should represent the discharge of the unitgraph applied to the largest rain. If the correspondence is poor, the initial coefficients are altered and another trial is made. By a series of converging approximations, the residual graph is made to correspond with the assumed distribution graph.

The procedure is set out below step by step and referred to the particular case of the River Rother at Woodhouse Mill, for which a unit hydrograph is derived from the storm of 14/15 May 1967. Phase A is standard procedure, Phase B is the Collins' approach.

Phase A: assembling and preparing the data

(1) Assemble all rainfall data available for the catchment under consideration and the storm period, including daily observations, recording rain-
gauge records and the synoptic weather maps of the region, if available.

(2) Derive a mean mass curve of rainfall for the catchment, for the period of
rain producing the hydrograph under study. Make provisional separation
of the rain into separate uniform periods.

*The catchment of the River Rother at Woodhouse Mill, Yorkshire is
shown in figure 7.18. The continuous rainfall record (mass curve) at*

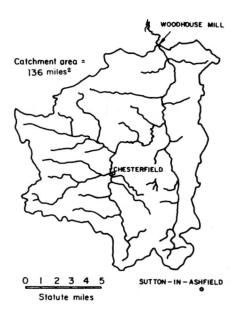

Figure 7.18 *Catchment of the River Rother at Woodhouse Mill*

*Sutton-in-Ashfield for the period of the storm is shown in figure 7.19.
The dashed line superimposed on the mass curve represents the idealised
intensities used and plotted in the appropriate time period of figure 7.20.
The total rainfall depth is measured daily at Chesterfield in the centre of
the catchment. The total recorded at Chesterfield on 15 May at 9.00 a.m.
was 1.42 in., as against 1.43 in. at Sutton. The precipitation was frontal,
widespread and comparatively uniform and so the recorded rainfall at
Sutton has been assumed catchment-wide.*

(3) Plot the hydrograph of total flow and separate baseflow (section 7.3.).
*In the hydrograph for the Rother at Woodhouse Mill (figure 7.20) the
position of the baseflow separation line was very indeterminate and so*

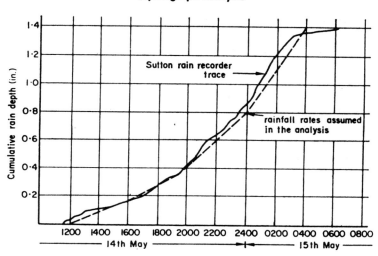

Figure 7.19 *Rainstorm of 14/15 May 1967, at Sutton-in-Ashfield*

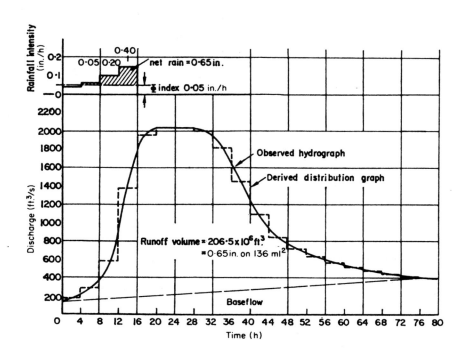

Figure 7.20 *Hydrograph of storm runoff and raingraph for River Rother at Woodhouse Mill, Yorkshire, 14–17 May 1967*

the end of surface runoff (the point N) was made about 2 days after the peak.

(4) Decide upon a unit period. As a general rule this should be not greater than one-quarter of the time from start of runoff to peak. Consider the provisional rain graph (from (2), page 168).

The choice of 4 h in this analysis was suitable for both rain and runoff.

(5) Consider the soil-moisture deficit of the catchment and use the Φ-index or f_{av} method for estimating storm loss for each rain period. If antecedent precipitation indices have been kept for the catchment, use these. Make an estimate of initial and subsequent loss rates. Compare the net rain derived in this way with the surface runoff expressed as depth on the catchment. If the two do not agree, one or the other must be adjusted.

The Rother catchment was wet, but not saturated, from light rains the day previous to the storm. It seemed reasonable to assume the first period's rain was lost entirely in making good the remaining soil-moisture deficit. The index was subsequently chosen to balance the net rain with surface runoff.

Phase B: using the data to derive the unitgraph

(6) Tabulate the relevant data in columns (1)–(9) of a table similar to table 7.4 and provide columns for the number of unit periods in the unit hydrograph under 'Distribution coefficients'.

The number of columns in table 7.4 under 'Distribution coefficients' is 16. This is the number of unit periods from the beginning of the last rain period to the end of surface runoff.

(7) Assume distribution coefficients of the unitgraph (representing percentage of total runoff per unit period) and arrange in appropriate columns.

(8) Determine the discharge, which, flowing constantly for one unit period would just equal 1 in. of net rain on the catchment.

This figure is found in this case to be

$$\frac{136 \times 27.9 \times 10^6}{12 \times 4 \times 3600} = 21960 \ ft.^3/s$$

(9) The first net rain is multiplied by this discharge and the product is distributed in percentages across the distribution coefficient columns by being multiplied in turn by each percentage coefficient. The various numbers are entered in the columns diagonally.

In this case $0.05 \times 21960 = 1098$, so in the first column $0.05 \times 1098 = 55$ and so on. Note that 55 was entered opposite the corresponding rain and not on the top row, which is ignored.

(10) The procedure of (9), above, is repeated for all net rains except the largest for which a dash is entered throughout.

It is purely fortuitous that the largest rain is at the end in this case.

TABLE 7.4 Derivation of unit hydrograph from a multi-period storm

(1) Day	(2) Time	(3) Period no.	(4) Rain (in.)	(5) Losses	(6) Net rain (in.)	(7) Average age	(8) Base-flow	(9) Net Q	Distribution Coefficients (per cent) 5.0	10.0	14.0	15.0	13.5	11.0	8.5	6.2	4.0	3.3	2.6	2.1	1.7	1.4	1.0	0.7	(11) Σ	(12) Residuals	(13) per cent
14 May	1200		0.16	0.20	0	180	145	35																			
	2400	1	0.25	0.20	0.05	290	160	130	55																	55	75	(0.8)
		2	0.40	0.20	0.20	580	175	405	220	110																330	75	(0.8)
15 May		3	0.60	0.20	0.40	1380	185	1195	—	440	154														594	601	6.9
		4				1950	200	1750		—	615	165														780	970	11.1
		5				2030	215	1815			—	660	148												808	1007	11.5
		6				2030	225	1805				—	594	121											715	1090	12.5
		7				1995	240	1755					—	484	93										577	1178	13.4
	2400	8				1820	255	1565						—	374	68									442	1123	12.8
16 May		9				1450	270	1180							—	273	44								317	863	9.8
		10				1095	280	815								—	176	36							212	603	6.9
		11				840	300	540									—	145	29						174	366	4.2
		12				720	315	405										—	114	24					138	267	3.1
		13				640	325	315											—	97	19				116	199	2.3
	2400	14				570	340	230												—	75	15			90	140	1.6
17 May		15				520	350	170													—	62	11		73	97	1.1
		16				480	365	115														—	44	8	52	63	0.7
		17				440	380	60															—	31	31	29	0.3
		18				420	395	25																—	0	25	0.3
2nd trial coefficients									6.4	10.9	12.6	13.5	13.4	12.5	9.4	6.7	4.1	3.2	2.4	1.8	1.3	1.0	0.6	0.4			
3rd trial coefficients									7.2	11.5	12.3	12.8	12.9	12.3	9.5	6.7	4.2	3.2	2.5	1.8	1.3	1.0	0.5	0.3			
Accepted coefficients									7.0	11.6	12.5	12.9	12.7	12.0	9.6	6.8	4.3	3.2	2.5	1.8	1.3	1.0	0.5	0.3			100.4

(11) The various discharges are now summed horizontally and entered in the Σ column.

(12) The Σ column discharge totals are now subtracted from column (9) and the remainders entered in the 'Residuals' column. These residuals are then converted into percentages of the unit distribution graph by dividing by the discharge of (8) (see page 170) multiplied by the largest rain, and subsequently multiplying by 100. The sum of the percentage column should be 100. The percentages that cannot have been influenced by the largest rainfall are bracketed and redistributed over the other coefficients so that the total, 100 per cent, remains constant. These percentages represent the distribution that would be necessary for the largest rain to make good the net Q of column (9). If they are the same as the assumed distribution coefficients then the unit distribution graph has been determined.

The residual 1090 of period 6 for example is converted thus

$$\frac{1090}{0.40 \times 21960} \times 100 = 12.5 \text{ per cent}$$

The percentages total comes to 100.4 because of rounding off. The bracketed figures have not been redistributed since the next trial requires fairly substantial changes in any case.

(13) If the differences between the trial coefficients and the adjusted (after redistribution) coefficients are large, then a new trial set must be adopted and steps (9)–(12) repeated until the differences are sufficiently small to be ignored (say <1 per cent). A weighted average of the previous trial and resulting adjusted coefficients should be used, as follows.

If P = the sum of residuals
Q = the sum of discharges of all the periods during which the largest rainfall would have been contributing
C_1 = the trial coefficient
C_2 = the calculated and adjusted coefficient
C_3 = the proposed new trial coefficient
then

$$C_3 = \frac{QC_1 + PC_2}{Q + P}$$

Alternatively, the weighting is often simply done in proportion to the total rain depth, the trial coefficients having the weighting of rain depths actually used, and the computed coefficients that of the largest rainfall.

(14) It is always wise to plot the distribution graph before deciding on acceptable coefficients. It may be found that small adjustments can help to give a smooth curve for the adopted unitgraph.

In the illustrated case three trials were made with calculations and a final adjustment (without recalculation) made after plotting the distribution graph and hence deriving the unitgraph shown in figure 7.21.

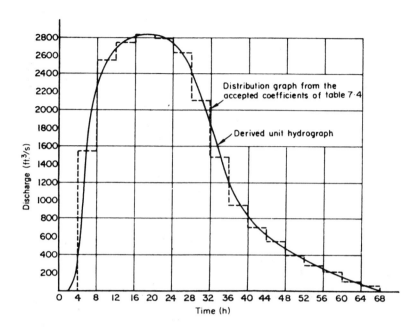

Figure 7.21 *4-h unit hydrograph for River Rother at Woodhouse Mill derived from the multi-period storm of figure 7.19*

The method is particularly useful when the largest rainfall is very great compared with the others since rapid convergence of the coefficients then takes place. This was not the case in the example illustrated. Too much reliance should not be placed on unitgraphs derived in this way until they have been used in practice and/or derived from a series of different storms, since the loss rates chosen have a critical influence on the resulting unitgraphs.

7.10 The instantaneous unit hydrograph

An extension of unitgraph theory is the concept of the instantaneous unit hydrograph or IUH. The IUH is the hydrograph of runoff from the instantaneous application of unit effective rain on a catchment.

Referring to figure 7.13 of section 7.6, the S-curve was seen to be simple method of deriving a unitgraph of period T h from the unitgraph of any other period t, by drawing two t h S-curves, T (h) apart. This is expressed in the equation

$$U(T, t) = \frac{t}{T} (S_t - S_{t-T}) \tag{7.2}$$

where $U(T,t)$ represents the ordinates of the T-h unitgraph derived from those of the t-h unitgraph. Now as T, progressively diminishing to zero, approaches dt, the right-hand side of equation 7.2 approaches the S-curve derivation, as can be seen graphically in figure 7.22. In equation form this is

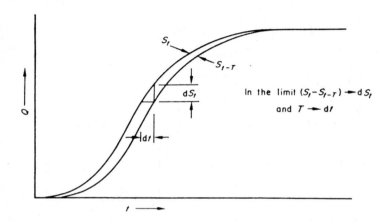

Figure 7.22   *The instantaneous unit hydrograph as the S-curve derivative*

$$U(0, t) = \frac{d(S_t)}{dt}$$

that is, the ordinate of the IUH at any time t is given by dS_t/dt at time t.

The IUH is a unique demonstration of a particular catchment's response to rain, independent of duration, just as the unitgraph is its response to rain of a particular duration. Since it is not time-dependent, the IUH is thus a graphical expression of the integration of all the catchment parameters of length, shape, slope condition etc. that control such a response.

The conversion of an IUH to a unitgraph of finite period is simple. The ordinate of an n-h unitgraph at time t is the average ordinate of the IUH for n h before t. From figure 7.23, it can be seen that the IUH is divided into n-h time intervals, and the averages of the ordinates at the beginning and end of each interval are plotted at the end of the intervals to provide the n-h unitgraph.

The IUH can be used to derive unitgraphs by flood routing as explained in section 8.6.

7.11 Synthetic unit hydrographs

In preceding sections it has always been assumed that some records have been available for the derivation of the unitgraph, but there are many catchments for

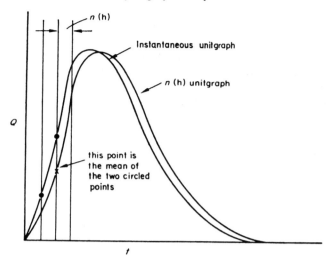

Figure 7.23 *The n* (h) *unitgraph derived from the IUH*

which there are no runoff records at all and for which unitgraphs may be required. In these circumstances hydrographs can be synthesised on the basis of past experience in other areas and applied as first approximations to the unrecorded catchment. Such devices are called *synthetic unitgraphs.*

The original approach is due to Snyder [6] who selected the three parameters of *hydrograph base width, peak discharge* and *basin lag* as being sufficient to define the unit hydrograph. These are shown in figure 7.24.

Snyder considered the catchment characteristics likely to affect unit hydrograph shape as being catchment area, shape of basin, topography, channel slopes, stream density and channel storage. He eliminated all these parameters except the first two by including them in a coefficient C_t. He dealt with the size and shape of catchment by measuring the length of the main stream channel and he proposed that

$$t_p = C_t(L_{ca}L)^{0.3}$$

where t_p = basin lag in h (that is, the time between mass centre of unit rain of t_r h duration and runoff peak flow.

L_{ca} = distance from gauging station to centroid of catchment area, measured along the main stream channel to the nearest point, in miles.

L = distance from station to catchment boundary measured along the main stream channel, in miles.

C_t = a coefficient depending on units and drainage basin characteristics and varying between 1.8 and 2.2 for the Appalachian Highlands catchments studied.

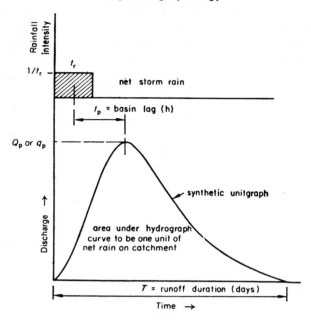

Figure 7.24 *Synthetic unitgraph parameters*

The equation for peak flow (per square mile of area) of the t_r unit graph was given by

$$q_p = C_p \cdot \frac{640}{t_p} \qquad (7.3)$$

where q_p is in cubic feet per second, and C_p is a coefficient depending on units and basin characteristics that varies between 0.56 and 0.69 for the Appalachian catchments and generally approaches its largest value as C_t approaches its lowest and vice versa.

Therefore the peak flow for the unitgraph is

$$Q_p = C_p \cdot \frac{640A}{t_p}$$

where A = catchment area in square miles.

The duration of surface runoff, or unit hydrograph baselength, T was given by Snyder by the empirical expression

$$T = 3 + 3\left(\frac{t_p}{24}\right)$$

where T is in days and t_p in hours. This expression gives a minimum baselength of 3 days for even small areas, a period much in excess of delay attributable to channel storage.

Snyder comments on this as being due to the 'subsurface storm flow', which has been defined by Hursh [7] as 'that portion of the storm flow which infiltrates into the surface soil but moves away from the area through the upper soil-horizons at a rate much in excess of normal groundwater seepage'. This is what is referred to in recent literature as *interflow* and for most practical purposes it is regarded as surface runoff.

t_r, the unit rain period, was assumed to equal $t_p/5.5$ in the study, since it was necessary to choose a single standard in all the catchment for the derivation of the formulae. This particular value was selected to make the unit of time equal to a minimum value below which further decrease would have little or no effect on basin lag or unitgraph peak discharge. If the actual length of the storm is not equal to t_r, but is t_R, equation 7.3 becomes

$$q_{pR} = C_p \cdot \frac{640}{t_p + (t_R - t_r)/4}$$

where q_{pR} = peak discharge (per square mile) of the t_R unitgraph, which allows for the generally observed reduction in unitgraph peak flows with longer periods of rain. Snyder proposed subsequently [8] an expression to allow for some variation in basin lag with variation in effective rainfall duration

$$t_{pR} = t_p + (t_R - t_r)/4$$

where t_{pR} = basin lag for a storm of duration t_R.

Linsley subsequently presented data [9] based on a study of Californian catchments and suggested modifications of Snyder's formulae and gave values of the various coefficients, as follows.

Basin lag: a new factor t_{p0} was introduced—the basin lag of an instantaneous storm—and used to derive t_{pR}, which has the same meaning as before.

$$t_{p0} = C_t(L_{ca}L)^{0.3} \text{ with } C_t \text{ (average)} = 0.5$$

$$t_{pR} = t_{p0} + (C_8 - 0.5)t_R$$

where C_8 (average) = 0.85.

Unitgraph peak flow (per square mile): $q_{pR} = C_p \cdot 640/t_{pR}$, where C_p varies between 0.35 and 0.50, and $Q_{pR} = q_{pR}A$

Time base of unitgraph: $T = 3 + (3 t_{pR}/24)$ days

The degree of divergence of the coefficients is an indication of how important it is to attempt to obtain some actual data on the lag in an ungauged catchment, and thus enable the value of C_t to be used with reasonable assurance.

A further advance in the subject was made by Taylor and Schwarz [10] in a study based on 20 drainage basins from 20 to 1600 square miles in area, in the north and central Atlantic states of the USA. They used the parameters L, L_{ca} as before and also introduced the slope of the main watercourse by defining

S_{st} as the slope of a uniform channel having the same length as the longest watercourse and an equal time of travel. The equations derived are as follows.

Basin lag: $t_{pR} = C'e^{m'}t_R$

where t_{pR} = lag in h from centroid of net rain to hydrograph peak
$\quad t_R$ = time in h from beginning to end of net rain
$\quad e$ = 2.7183
$\quad m'$ = rate of change of lag with storm duration
$\quad C'$ = lag of instantaneous unit hydrograph
m' and C' are derived from the following equations

$$m' = 0.212/(LL_{ca})^{0.36}$$
$$C' = 0.6/\sqrt{(S_{st})}$$

where L and L_{ca} have the same definitions as previously and

$$S_{st} = \left[\frac{n}{(1/S_1^{\frac{1}{2}} + 1/S_2^{\frac{1}{2}} + \ldots + 1/S_n^{\frac{1}{2}})}\right]^2$$

where n = Manning's coefficient of roughness for the natural watercourse

S_1, S_2 etc. = the slopes of individual sections, of equal length, into which the main watercourse can be conveniently divided.

Peak discharge (per square mile) of unitgraph: $q_{pR} = C''e^{m''}t_R$

where

$$C'' = 382(LL_{ca})^{-0.36}$$
$$m'' = 0.121S_{st}^{0.142} - 0.05$$

Base width of unitgraph: $T = 5(t_{pR} + t_R/2)$ h

The authors give a nomogram in their paper for the solution of equations and make a number of observations about the use of their method. These include the suggestions that major tributaries should be treated separately, and that the equations should be limited to the results of moderate and major storms of uniform distribution over geographical areas similar to those from which they were derived. The equations are given here as they form a useful addition to the literature on synthesis of unitgraphs.

Unit hydrographs can also be synthesised by the methods of stream-flow routing and section 8.6 describes such a technique. It has been placed in chapter 8 because a knowledge of routing procedure is essential to its understanding.

7.12 Synthetic unit hydrograph from catchment characteristics by the FSR method

The FSR provides a method for synthesising a 1-h unit hydrograph for an ungauged catchment and for choosing an appropriate design rain to apply to it [1].

The unit hydrograph is based on three parameters similar to those used by previous investigators: time to peak, peak discharge and hydrograph base width. However, the definitions are not quite the same.

The time to peak T_p (h) is the time to peak of a 1-h unit hydrograph measured from the *start of response runoff*, and is given by

$$T_p = 46.6 \, (MSL)^{0.14} \, (S1085)^{-0.38} \, (1 + URBAN)^{-1.99} \, (RSMD)^{-0.4} \quad (7.4)$$

where MSL   is the main stream length (km), measured from the 1:25 000 map (a detailed procedure for obtaining MSL is given in reference 1 (I.4.2.2.)).

S1085   is the slope obtained by identifying two points at distance of 10 per cent and 85 per cent respectively of the main stream length from the catchment outfall as shown on 1:25 000 maps and determining the difference in their elevations and the stream length between them. This slope is defined in m/km.

URBAN is the fraction of catchment in urban development.

RSMD   is the 1-day M5 rainfall, less effective mean soil moisture deficit. The value of RSMD for any point in Britain and Ireland can be read from figures 7.25, 7.26 and 7.27 reproduced from reference 11. Alternatively it can be calculated as described in example 7.2.

If any records of rainfall and response runoff exist, a more reliable estimate can be made from

$$T_p = 0.9 \, LAG$$

where LAG is time (h) from the centroid of the rain profile to the peak runoff, or to a 'centroid of peaks' if there is more than one.

The peak of the unit hydrograph Q_p in m³/s per 100 km² is estimated from

$$Q_p = 220/T_p \text{ and its time base TB} = 2.52 T_p$$

These three parameters T_p, Q_p and TB enable a triangular unit hydrograph to be drawn. This is the 1-h unit hydrograph. It is illustrated in figure 7.30.

The duration D (h) of a design storm depends on T_p and the mean annual rainfall (SAAR), and is given by

$$D = (1 + SAAR/1000) \, T_p \quad (7.5)$$

SAAR can be obtained for any point in the British Isles from appendix A.

The rainfall intensity to be chosen depends on the return period of the design flood. The return period is the time that, on average, elapses between two

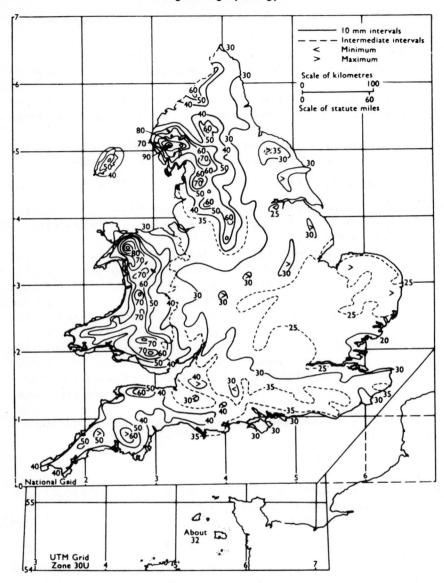

Figure 7.25 *RSMD (mm) for England and Wales*

events that equal or exceed a particular level (see also section 9.3.2); the return periods of floods and their causative rainfall are not necessarily the same, since there are other factors involved. However, the FSR provides a relationship for a recommended storm return period to yield a flood peak of required return period by the design method described, from which figure 7.28 has been derived.

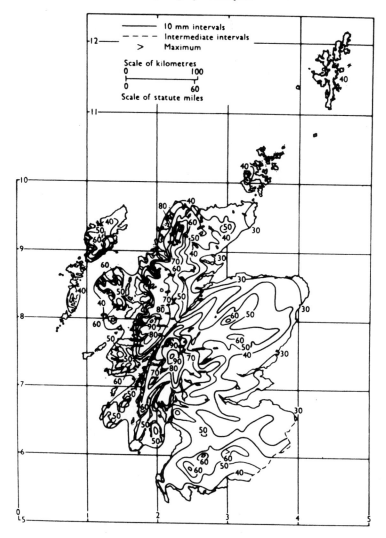

Figure 7.26   *RSMD (mm) for Scotland*

Having established the required return period of rainfall of duration D, the total storm rainfall over the particular catchment can be obtained as described in section 2.9.4 and this total distributed in time by the appropriate profile (section 2.9.5). An example will illustrate the details of the whole process.

Example 7.2.   By means of a synthetic unit hydrograph, estimate the flood peak with a 200-year return period for the River Brathay at Skelwith Force at NGR NY341032 (figure 7.29).

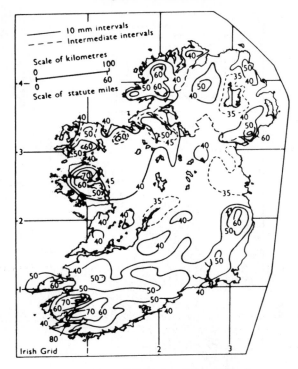

Figure 7.27 *RSMD (mm) for Ireland*

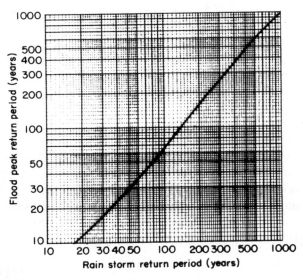

Figure 7.28 *Recommended storm return period to yield a flood peak of required return period by the design method*

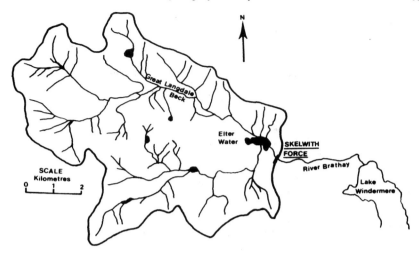

Figure 7.29 *The catchment area of the River Brathay at Skelwith Force*

(1) Outline the catchment area on the 1:25 000 map and measure the area (AREA) and length of the main stream MSL:
AREA = 50.4 km^2
MSL   = 12.2 km

(2) Channel slope (S1085) is the average slope in m/km between two points at 10 per cent and 85 per cent of the main stream length measured from the outlet. Since

elevation at 85 per cent point = 137 m

and

elevation at 10 per cent point = 58 m

therefore

S1085 = 8.6 m/km

(3) The average annual rainfall (SAAR) is obtained by taking a weighted average over the catchment:

SAAR = 2500 mm

(4) An estimate of RSMD can be made by reference to figure 7.25. Alternatively it can be calculated by finding

2-day M5 rainfall (average for catchment)	= 135 mm
r (average for catchment)	= 15 per cent

and so determining (table 2.9)

$$24\text{-h M5} = 0.75 \times 135 \qquad\qquad = 101 \text{ mm}$$

Converting this to 1-day M5 (table 2.10)

$$101/1.11 \qquad\qquad = 91.2 \text{ mm}$$

and knowing

$$\text{ARF (table 2.8)} \qquad\qquad = 0.95$$

gives 1-day M5 catchment rainfall $\qquad = 86.6$ mm
Finally deducting s.m.d. (figure 4.8) (= 4 mm) gives $\qquad 82.6$ mm

$$\text{RSMD} = 82.6 \text{ mm}$$

or rounding up

$$\text{RSMD} = 83 \text{ mm}$$

(5) There is no urban development, so

$$\text{URBAN} = 0$$

(6) T_p Q_p and TB can now be calculated from equation 7.4.

$$T_p = 46.6 \,(12.2)^{0.14} \,(8.6)^{-0.38} \,(83)^{-0.4} = 4.99$$

So that

$T_p = 5.0.$
$Q_p = 220/T_p/100 \text{ km}^2$
$\quad = 44 \text{ m}^3/\text{s}/100 \text{ km}^2$
$\quad = 22.2 \text{ m}^3/\text{s for } 50.4 \text{ km}^2$
$T_B = 2.52\, T_p = 12.6$

So take

$T_B = 13$ h.

(7) The basic data interval $T \simeq T_p/5$. Hence $T = 1$ h.

(8) The design storm duration can now be calculated from equation 7.5:

$$D = (1 + \text{SAAR}/1000)T_p$$
$$= (1 + 2500/1000)5$$
$$= 17.5$$

It is convenient to make D an odd integer multiple of T.

Hence

$$D = 17 \text{ h}.$$

(9) It is now necessary to decide the return period of storm that will produce the appropriate return period peak flow. For a 200-year peak flow the recommended return period is obtained from figure 7.28: the storm return period is 240 years.

(10) A rainstorm of duration 17 h, having a return period of 240 years, can be found by first consulting table 2.9, using the value of r found in step (4) above and a duration of 17 h

$$r = 15 \text{ so that 17 h M5/2-day M5} = 62 \text{ per cent}$$

and

$$62 \text{ per cent of 2-day M5} = 0.62 \times 135 = 83.7$$

Hence take

$$17 \text{ h M5 as 84 mm.}$$

(11) It is now necessary to convert 17-h M5 to 17-h M240. From table 2.6, for M5 = 84 mm, by interpolation M240 requires a growth factor of 1.71, therefore

$$17\text{-h M240} = 84 \times 1.71 = 144 \text{ mm}$$

(12) This is a point rainfall value and must now be reduced to a catchment average using ARF (table 2.8). Interpolating

$$\text{ARF (for 17 h and 50 km}^2) = 0.95$$

Therefore

$$\text{rainfall } P = 137 \text{ mm}$$

(13) A value of catchment wetness index is now obtained using SAAR = 2500 mm in figure 4.9:

$$\text{CWI} = 127$$

(14) A percentage runoff figure is now needed. This is dependent on the fractions of the catchment covered by various classes of soil. (A definition of SOIL will be found in section 9.4.) Maps giving soil classifications for the British Isles are provided in appendix A, identified by the letters RP (runoff potential), in sections, together with rainfall maps.

When such maps are not available, judgement based on table 4.2, should be used. In this case the whole of the catchment is mountainous upland and is soil class 5. Then from section 9.4

$$SOIL = 0.5$$

Standard percentage runoff is calculated from

$$SPR = 95.5 \ SOIL + 0.12 \ URBAN$$

So

$$SPR = 47.7 \text{ per cent}$$

and the appropriate runoff percentage (PR) for the design event is obtained from

$$
\begin{aligned}
PR &= SPR + 0.22 \ (CWI - 125) + 0.1 \ (P - 10) \\
&= 47.7 + 0.22 \ (127 - 125) + 0.1 \ (127) \\
&= 60.8 \text{ per cent}
\end{aligned}
$$

Therefore net rain for application to the synthetic unit hydrograph = 137 x 0.608 = 84 mm

(15) This net rain must now be applied to the unit hydrograph in accordance with a 75 per cent winter storm profile. This distribution should take place over the 17-h storm period and so a stepped distribution graph over 17 one-hour periods is required. Each interval will be approximately 6 per cent of duration. The percentage rain is obtained from figure 2.17.

Duration (per cent)	6	18	30	42	54	66	78	90	100
Rain (per cent)	16	43	61	73	83	89	94	98	100
Increment rain (per cent)	16	27	18	12	10	6	5	4	2
Increment rain (mm)	13.4	22.7	15.1	10	8.4	5.0	4.2	3.4	1.6

Each of the increments, except the first (the peak), represents the total of two equal 1 h periods of rain. This rain is now arranged symmetrically about the centre line and applied to the synthetic unit hydrograph.

The stepped distribution graph of rain is shown in figure 7.30 and its application to the synthetic unitgraph is made in table 7.5.

7.13 The application of rain to unit hydrographs

7.13.1. General. In using unit hydrographs for hydrological forecasting, it is necessary to adjust the unitgraph to the time-span of the uniform rain. Hence t_1 hours of continuous uniform rain should be applied to a t_1-hour unitgraph

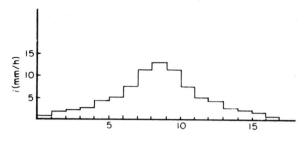

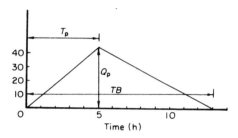

Figure 7.30 *Synthetic 1-h unitgraph and hyetograph for example 7.2*

and t_2 hours to a t_2-hour graph. Methods of converting a t_1 to a t_2 unitgraph were described earlier in this chapter. The only mention made so far, however, of how to choose a particular duration and intensity of rain to apply was in section 7.12 and example 7.2, where the FSR recommendations were used.

The operation of applying rain to unitgraphs is carried out for many different reasons. The design and construction of different projects may involve the estimation of many different floods. For example, the dam designer wants to make his spillways sufficiently large so that the safety of the dam is not threatened in its lifetime. The dam builder wants to know what risk he runs of various floods occurring in the 3 or 4 years he may be building in a river channel, so that his cofferdams and tunnels are economically sized. The flood protection engineer wishes to be sure his levees will not be overtopped more often than at some design frequency that he regards as economic; or perhaps they must not be overtopped at all if the safety of human life is concerned. The estuarial or river port manager wants to know the depth of water he can rely on in navigable channels related to a probability of availability, and so on.

In any or all of these cases it may be desirable to choose a design rain of particular frequency and magnitude and apply it to a catchment unitgraph as a means of flow prediction. This is particularly so because in many parts of the world there are reasonably long-term rainfall records though very sparse or short-term runoff records.

TABLE 7.5   Convolution of the 1-h unit hydrograph with net rain[a]

Hour	Net rain (cm)	Hour from origin and 1-h unit hydrograph ordinates × (area/100) (m³/s)																
		0	1	2	3	4	5	6	7	8	9	10	11	12	13	14	15	16 (h)
		0	4.44	8.87	13.31	17.74	22.18	19.40	16.63	13.86	11.09	8.32	5.54	2.77	0	0	0	0 (m³/s)
1	0.08	0	0.4	0.7	1.1	1.4	1.8	1.6	1.3	1.1	0.9	0.7	0.4	0.2				
2	0.17		0	0.8	1.5	2.3	3.0	3.8	3.3	2.8	2.4	1.9	1.4	0.9	0.5			
3	0.21			0	0.9	1.9	2.8	3.7	4.7	4.1	3.5	2.9	2.3	1.8	1.2	0.6		
4	0.25				0	1.1	2.2	3.3	4.4	5.6	4.8	4.2	3.5	2.8	2.1	1.4	0.7	
5	0.42					0	1.9	3.7	5.6	7.5	9.3	8.2	7.0	5.8	4.7	3.5	2.3	1.2
6	0.50						0	2.2	4.4	6.7	8.9	11.1	9.7	8.3	6.9	5.6	4.2	2.8
7	0.75							0	3.3	6.7	10.0	13.3	16.6	14.6	12.5	10.4	8.3	6.2
8	1.14								0	5.1	10.1	15.2	20.2	25.3	22.1	19.0	15.8	12.6
9	1.34									0	6.0	11.9	17.8	23.8	29.7	26.0	22.3	18.6
10	1.14										0	5.1	10.1	15.2	20.2	25.3	22.1	19.0
11	0.75											0	3.3	6.7	10.0	13.3	16.6	14.6
12	0.50												0	2.2	4.4	6.7	8.9	11.1
13	0.42													0	1.9	3.7	5.6	7.5
14	0.25														0	1.1	2.2	3.3
15	0.21															0	0.9	1.9
16	0.17																0	0.7
17	0.08																	0
	8.38	0	0.4	1.5	3.5	6.7	11.7	18.3	27.0	39.6	55.9	74.5	92.3	107.6	116.2	116.6	109.9	99.5

Base flow 3.3

119.9

Peak flow of the flood of 200-year return period is 120 m³/s

[a]Each increment is 1-h rain is multiplied in turn by each hourly ordinate of the unit hydrograph, successive products being moved successively 1 h to the right.
The table may be continued to the right to provide further ordinates of the total hydrograph.
Baseflow is calculated from equation 7.1

$$\text{ANSF} = (3.26 \times 10^{-4})(CWI - 125) + (7.4 \times 10^{-4})\,RSMD + (3 \times 10^{-3})$$

Since CWI = 127 and RSMD = 83, therefore

$$\text{ANSF} = (621 \times 10^{-4}) + 0.003$$
$$= 0.0651 \text{ m}^3/\text{s/km}^2$$

Hence

baseflow = 50.4 × 0.0651
= 3.3 m³/s

The rainfall records must therefore be examined and frequency analyses made of the incidence of particular 24-h rainfall depths. Rainfall observations are comparatively rarely made more often than this, although clearly very intense storms of rain frequently last for shorter periods. Much depends on catchment size. If it is very large, then 24-h frequencies may be adequate for choosing design rain. Most often, however, shorter periods of rain at greater intensity will represent the greatest flood danger and efforts must be made to find shorter interval data, or to start measuring it by autographic recorder as soon as project investigation commences. Extrapolations can be made by recording the frequencies with which certain rainfall depths occur in 96-h, 72-h, 48-h and 24-h periods. Many data are already available from sources in the USA [12-16], which though directed obviously to catchments in the USA, are applicable with care elsewhere. In Britain, reference should be made to the Flood Studies Report [17]. The Meteorological Office contribution to this Report allows estimates of expected rainfall amounts to be made for any place in the United Kingdom with a given return period for durations from 15 s to 30 days, as discussed in chapter 2. Reference may also be made to Bleasdale [18], to *British Rainfall 1939 (et seq.)* and to Wiesner [19].

The process of using the unit hydrograph therefore involves the following steps.

(a) An examination of all relevant data (including frequency analyses): in particular, any autographic rain recordings, and a subsequent selection of a total precipitation that could, in the judgement of the individual concerned, occur in a reasonably short time period, over the catchment concerned. (Note that where the catchment is large and it seems unlikely that uniform precipitation could or would occur, it should be split into a series of sub-catchments and each of these should be assigned a precipitation.

(b) From the total precipitation, thus chosen, all losses must be subtracted. These will include interception, infiltration (having regard to the soil-moisture deficiency) and evapotranspiration.

(c) The resultant net (or effective) rainfall is then applied to a unitgraph adjusted to the correct time base, and the resulting storm hydrograph ordinates obtained. If more than one hydrograph is concerned the temporal spacing of the rain must be decided, usually, but not necessarily, to obtain the maximum possible combination of ordinates.

(d) Baseflow, assessed separately, is now added to give the total storm runoff.

The choice of the s.m.d. on which a design storm may fall is significant. It is not always reasonable to presuppose a saturated catchment, and a hypothetical infiltration capacity curve, a Φ-index or coaxial correlation graph (for example figure 4.7) can be used with this in mind. For catchments in the British Isles, when no other information is available, the mean value of s.m.d. from figure 4.8 should be used.

7.13.2 The 1-h unit hydrograph

In recent years the concept of design rainfall has changed to simulate natural events more precisely. Rain does not fall naturally at uniform intensity for long periods. With computers now in common use, the simplications that were previously desirable are no longer necessary since the tedious calculations involved need no longer be performed manually.

Rainfall events are now specified as total depths (mm) falling with particular durations and to particular storm profiles, the profile being compared in terms of their peakedness and seasonality. Thus, the United Kingdom 75 per cent winter profile is a storm profile that is typical of storm rainfall from October to March, or the winter season, in Britain, and is described in section 2.9.5.

For design purposes, generally large events are postulated. Rainfall durations thus tend to be of many hours length. Accordingly, it is frequently convenient to use a 1-h rain increment applied to a 1-h unit graph, and to convolute the net rain profile with the hourly ordinates of the unitgraph, as was done in example 7.2.

References

1. Natural Environment Research Council. *Flood Studies Report*, Vol. I, NERC, 1975, Chapter 6
2. SHERMAN, L. K. Stream flow from rainfall by the unitgraph method. *Eng. News Record*, **108**, (1932) 501
3. BERNARD, M. An approach to determinate stream flow. *Trans. Am. Soc. Civ. Eng.*, **100**, (1935) 347
4. LINSLEY, R. K., KOHLER, M. A. and PAULHUS, J. L. H. *Applied Hydrology*, McGraw-Hill, New York, 1949, pp. 448–9
5. COLLINS, W. T. Runoff distribution graphs from precipitation occurring in more than one time unit. *Civ. Eng.*, **9**, No. 9 (September 1939) 559
6. SNYDER, F. F. Synthetic unitgraphs. *Trans. Am. Geophys. Union, 19th Annual Meeting 1938*, Part 2, p. 447
7. HURSH, C. R. Discussion on Report of the committee on absorption and transpiration. *Trans. Am. Geophys. Union, 17th Annual Meeting 1936*, p. 296
8. SNYDER, F. F. Discussion in ref. 6
9. LINSLEY, R. K. Application of the synthetic unitgraph in the western mountain States. *Trans. Am. Geophys. Union, 24th Annual Meeting 1943*, Part 2, p. 580
10. TAYLOR, A. B.and SCHWARZ, H. E. Unit hydrograph lag and peak flow related to basin characteristics. *Trans. Am. Geophys. Union*, 33, (1952) 235
11. Institution of Civil Engineers. *Floods and Reservoir Safety: an engineering guide*, I.C.E., London, 1978
12. PAULHUS, J. L. H. and GILMAN, C. S. Evaluation of probable maximum precipitation. *Trans. Am. Geophys. Union*, 34, (1953) 701
13. Generalised estimates of probable maximum precipitation over the U.S. east of the 105th meridian. *Hydrometeorological Report No. 23, U.S. Weather Bureau*, Washington D.C., 1947

14. Generalised estimates of probable maximum precipitation of the United States west of the 105th meridian for areas to 400 square miles and durations to 24 hours. *Tech. Paper 38, U.S. Weather Bureau*, Washington D.C., 1960

15. Manual for depth duration area analysis of storm precipitation. *U.S. Weather Bureau Co-operative Studies Tech. Paper No. 1*, Washington D.C., 1946

16. HERSHFIELD, D. M. Estimating the probable maximum precipitation. *Proc. Am. Soc. Civ. Eng.*, 87, (September 1961) 99

17. National Environment Research Council. *Flood Studies Report*, Vol. II, NERC, 1975

18. BLEASDALE, A. The distribution of exceptionally heavy daily falls of rain in the United Kingdom. *J. Inst. Water Eng.*, 17, (February 1963) 45

19. WIESNER, C. J. Analysis of Australian storms for depth, duration, area data. *Rain Seminar, Commonwealth Bureau of Meteorology, Melbourne*, 1960

Further reading

BARNES, B. S. Consistency in unitgraphs. *Proc. Am. Soc. Civ. Eng.*, 85, HY8 (August 1959) 39

BUIL, J. A. Unitgraphs for non uniform rainfall distribution. *Proc. Am. Soc. Civ. Eng.*, 94, HY1 (January 1965) 235

MORGAN, P. E. and JOHNSON, S. M. Analysis of synthetic unitgraph methods. *Proc. Am. Soc. Civ. Eng.*, 88, HY5 (September 1962) 199

MORRIS, W. V. Conversion of storm rainfall to runoff. *Proc. Symposium No. 1, Spillway Design Floods*, N.R.C., Ottawa, 1961, p. 172

Problems

7.1 A catchment area is undergoing a prolonged rainless period. The discharge of the stream draining it is $100 \ m^3/s$ after 10 days without rain, and $50 \ m^3/s$ after 40 days without rain. Derive the equation of the depletion curve and estimate the discharge after 120 days without rain.

7.2 A catchment is suffering from a drought. The discharge of the stream draining it is $75 \ m^3/s$ after 12 days without rain and $25 \ m^3/s$ after 40 days without rain. Derive the equation of the stream's depletion curve and estimate the discharge 60 days into the drought.

7.3 The recession limb of a hydrograph, listed below, is to be divided into runoff and baseflow. Carry out this separation

(a) by finding the point of discontinuity on the recession limb,
(b) by finding the depletion curve equation and extrapolating back in time.

Comment on your results.

Time (h)	Flow (m³/s)	Time (h)	Flow (m³/s)
15	41.1	33	10.0
18	35.8	36	8.3
21	25.0	39	7.0
24	19.2	42	5.8
27	15.1	45	4.9
30	12.2	48	4.1

7.4 Describe how to derive a master depletion curve for a river. What would you use it for and why?

7.5 A drought is ended over a catchment area of 100 km² by uniform rain of 36 mm falling for 6 h. The relevant hydrograph of the river draining the area is given below, the rain period having been between hours 3 and 9. Use these data to predict the maximum discharge that might be expected following a 50 mm fall in 3 h on the catchment. Qualify the forecast appropriately.

Hours	Discharge (m³/s)	Hours	Discharge (m³/s)
0	3	24	25
3	3	27	21
6	10	30	17
9	25	33	13.5
12	39	36	10.5
15	43	39	8
18	37	42	5.5
21	30.5	45	4
		48	3.9

7.6 Describe how you would proceed to separate baseflow from the hydrograph of a stream's discharge.

7.7 Write down the three major principles of unit hydrograph theory illustrating their application with sketches.

  Given below are three unit hydrographs (all values in ft.³/s) derived from separate storms on a small catchment, all of which are believed to have resulted from 3-h rains. Derive the average unit hydrograph and confirm its validity if the drainage area is 5.25 square miles.

Hours	Storm 1	Storm 2	Storm 3
0	0	0	0
1	165	37	25

2	547	187	87
3	750	537	260
4	585	697	505
5	465	608	660
6	352	457	600
7	262	330	427
8	195	255	322
9	143	195	248
10	97	135	183
11	60	90	135
12	33	52	90
13	15	30	53
14	7	12	24
15	0	0	0

7.8 Listed below are three hydrographs derived from three separate uniform-intensity storms each lasting 3 h. The gross rainfall for storm A was 14 mm, for storm B was 24 mm, and for storm C was 19 mm. The Φ-index for the catchment is estimated to be $1\frac{1}{3}$ mm/h. Derive the average unit hydrograph for the catchment and confirm its validity if the drainage area is 13.60 km². All values are in m³/s.

Hours	Storm A	Storm B	Storm C
0	0	0	0
1	1.84	0.82	0.42
2	6.10	4.16	1.46
3	8.36	11.98	4.35
4	6.52	15.54	8.45
5	5.18	13.56	11.04
6	3.92	10.20	10.04
7	2.92	7.36	7.14
8	2.17	5.68	5.39
9	1.59	4.34	4.16
10	1.08	3.02	3.06
11	0.67	2.00	2.27
12	0.37	1.16	1.50
13	0.17	0.66	0.89
14	0.08	0.26	0.41
15	0	0	0

7.9 The 3-h unit hydrograph derived from a catchment of 14.5 km^2 is given below

Hours	Unit hydrograph (m^3/s)	Hours	Unit hydrograph (m^3/s)
0	0	8	3.6
1	0.3	9	2.8
2	1.0	10	2.0
3	2.9	11	1.5
4	5.6	12	1.0
5	7.3	13	0.6
6	6.7	14	0.3
7	4.8	15	0

What peak discharge would be expected from a 4-h rainfall at a uniform intensity of 15 mm/h, followed immediately by a 3-h storm at a uniform intensity of 10 mm/h? Assume a constant storm loss of 3 mm/h and a baseflow starting at 1.2 m^3/s at hour 0 and rising at 0.1 m^3/s per hour until after the peak.

7.10 The hydrograph tabulated below was observed for a river draining a 40 square miles catchment, following a storm lasting 3 h

Hour	ft.^{3}/s	Hour	ft.^{3}/s	Hour	ft.^{3}/s
0	450	24	3500	48	1070
3	5500	27	3000	51	950
6	9000	30	2600	54	840
9	7500	33	2210	57	750
12	6500	36	1890	60	660
15	5600	39	1620	63	590
18	4800	42	1400	66	540
21	4100	45	1220		

Separate baseflow from runoff and calculate total runoff volume. What was the net rainfall in inches per hour? Comment on the severity and likely frequency of such a storm in the United Kingdom.

7.11 The 4-h unit hydrograph for a river-gauging station draining a catchment area of 554 km^2, is given below.

Time (h)	Unit hydrograph (m³/s)	Time (h)	Unit hydrograph (m³/s)
0	0	11	76
1	11	12	62
2	60	13	51
3	120	14	39
4	170	15	31
5	198	16	23
6	184	17	16
7	153	18	11
8	127	19	6
9	107	20	3
10	91	21	0

Make any checks possible on the validity of the unitgraph. Find the probable peak discharge in the river, at the station from a storm covering the catchment and consisting of two consecutive 3-h periods of net rain of intensities 12 and 6 mm/h respectively. Assume baseflow rises linearly during the period of runoff from 30 to 70 m³/s.

7.12 The 3-h unit hydrograph for a river-gauging station draining an 835 km² catchment is listed below.

Time (h)	3-h unit hydrograph (m³/s)	Time (h)	3-h unit hydrograph (m³/s)
0	0	12	85
1	22	13	68
2	120	14	55
3	240	15	42
4	318	16	30
5	298	17	20
6	250	18	12
7	206	19	7
8	174	20	3
9	144	21	0
10	123		
11	102		

An intermittent storm lasting 7 h covers the catchment, the gross rainfall being: 17 mm/h for 4 h, followed by 12 mm/h for a further 3 h. The Φ-index for the catchment is 7 mm/h. Assuming baseflow is constant at 40 m³/s, provide an estimate of the maximum discharge and its time of occurrence from the start of the storm.

7.13 Using the data and catchment of 7.11 find the probable peak discharge in the river, at the station, from a storm covering the catchment and consisting of three consecutive 2-h periods of rain producing 7, 14 and 12 mm runoff respectively. Assume baseflow rises from $10 \, m^3/s$ to $20 \, m^3/s$ during the total period of runoff.

7.14 The 4-h unit hydrograph for a $550 \, km^2$ catchment is given below

Hours	Q (m^3/s)	Hours	Q (m^3/s)
0	0	11	76
1	11	12	62
2	71	13	51
3	124	14	40
4	170	15	31
5	198	16	27
6	172	17	17
7	147	18	11
8	127	19	5
9	107	20	3
10	90	21	0

A uniform-intensity storm of duration 4 h with an intensity of 6 mm/h is followed after a 2-h break by a further uniform-intensity storm of duration 2 h and an intensity of 11 mm/h. The rain loss is estimated at 1 mm/h on both storms. Baseflow was estimated to be $10 \, m^3/s$ at the beginning of the first storm and $40 \, m^3/s$ at the end of the runoff period of the second storm.

Compute the likely peak discharge and its time of occurrence.

7.15 Using a FSR (1975) synthetic unit hydrograph with $T_p = 8$ h for a catchment of $350 \, km^2$, deduce the surface runoff contribution to the discharge of the river draining the catchment at the end of the fourth hour of continuous rain of 10 mm/h, if the runoff coefficient is 70 per cent. Assume runoff starts at start of rain.

Define the parameters on which Snyder's original synthetic unit hydrograph was based and comment on the differences from the FSR synthetic unitgraph values.

7.16 A catchment of $76 \, km^2$ area drains to an outfall at National Grid Reference SS 742 480.

The main stream length is 15 km. Assume SAAR = 1750 mm and S1085 = 15.0 with no urban development.

Estimate the flood flow at the outfall with a 65 year return period, assuming that 50% of rainfall appears as runoff and ignoring baseflow.

7.17 (a) The 2 h unit hydrograph for a 370 km^2 catchment is listed below. Transform it to a 6 h unit graph and estimate the peak discharge from a 6 h storm of uniform intensity of 4 mm/h. Assume the rain loss amounts to 1mm/h and that baseflow is a constant 3 m^3/s. Is there any way you can check if the unit graph is valid?

Hours	Qm^3 (s)	Hours	Qm^3 (s)	Hours	Qm^3 (s)
0	0	12	65	24	18
2	8	14	54	26	13
4	22	16	45	28	9
6	44	18	37	30	5
8	67	20	29	32	2
10	72	22	23	34	0

(b) Comment on the assumption of constant baseflow, and describe a way in which baseflow at the end of a period of surface runoff might be estimated.

8 Flood Routing

8.1 Introduction

Civilisation has always developed along rivers, whose presence guaranteed access to and from the sea coast, irrigation for crops, water supplies for urban communities and latterly power development and industrial water supply. The many advantages have always been counterbalanced by the dangers of floods and, in the past, *levees* or flood banks were built along many major rivers to prevent inundation in the flood season. In more recent times storage reservoirs have been built as the principles of dam construction became better understood and other measures like relief channels, storage basins and channel improvements are continually under construction in many parts of the world. It is important for such works that estimates can be made of how the measures proposed will affect the behaviour of flood waves in rivers so that economic solutions can be found in particular cases. *Flood routing* is the description applied to this process. It is a procedure through which the variation of discharge with time at a point on a stream channel can be determined by consideration of similar data for a point upstream. In other words it is a process that shows how a flood wave can be reduced in magnitude and lengthened in time (*attenuated*) by the use of storage in the *reach* between the two points.

8.2 The storage equation

Since the methods of flood routing depend on a knowledge of storage in the reach, a way of evaluating this must be found. There are two ways of doing this. One is to make a detailed topographical and hydrographical survey of the river reach and the riparian land and so determine the storage capacity of the channel at different levels. The other is to use the records of past levels of flood waves at the two points and hence deduce the reach's storage capacity. It is assumed that such storage capacity will not change substantially in time and so may be used to route the passage of large and more critical, predicted floods. As many data as possible are required for the second method, which is the one generally used, including flow records at the beginning and end of the reach and on any tributary streams joining it, and rainfall records over any areas contributing direct runoff to it.

Storage in the reach of a river is divided into two parts, *prism* and *wedge* storage, which are illustrated in figure 8.1. This is simply because the slope of the surface is not uniform during floods (see section 6.2).

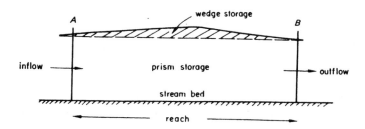

Figure 8.1   *Storage in a river reach*

If the continuity of flow through the reach shown in figure 8.1 is now considered, it is clear that what enters the reach at point A must emerge at point B, or temporarily move into storage.

$$I = D + \frac{dS}{dt}$$

where I = inflow to the reach
D = discharge from the reach
dS/dt = rate of change in reach storage with respect to time.

This equation is approximated, for a time interval t, by

$$\frac{I_1 + I_2}{2} t - \frac{D_1 + D_2}{2} t = S_2 - S_1 \qquad (8.1)$$

where subscripts 1 and 2 denote values at the beginning and end respectively of the time t. The time t is called the *routing period* and it must be chosen sufficiently short so that the assumption implicit in equation 8.1 (that is, the inflow and outflow hydrographs consist of a series of straight lines), does not depart too far from actuality. In particular, if t is too long it is possible to miss the peak of the inflow curve, so the period should be kept shorter than the travel time of the flood wave crest through the reach. On the other hand, the shorter the routing period the greater the amount of computation to be done.

8.3 Reservoir routing

If equation 8.1 is now arranged so that all known terms are on one side, the expression becomes

$$\tfrac{1}{2}(I_1 + I_2)t + (S_1 - \tfrac{1}{2}D_1 t) = (S_2 + \tfrac{1}{2}D_2 t) \qquad (8.2)$$

The routing process consists of inserting the known values to obtain $S_2 + \tfrac{1}{2}D_2 t$ and then deducing the corresponding value of D_2 from the relationship connecting storage and discharge. This method was first developed by L. G. Puls of the U.S.Army Corps of Engineers.

The simplest case is that of a reservoir receiving inflow at one end and discharging through a spillway at the other. In such a reservoir it is assumed there is no wedge storage and that the discharge is a function of the surface elevation, provided that the spillway arrangements are either free-overflow or gated with fixed gate openings. Reservoirs with sluices can be treated also as simple reservoirs if the sluices are opened to defined openings at specified surface-water levels, so that an elevation–discharge curve can be drawn. The other required data are the elevation–storage curve of the reservoir and the inflow hydrograph.

Example 8.1. An impounding reservoir enclosed by a dam has a surface area that varies with elevation as shown by the relationship of figure 8.2(a). The dam is equipped with two circular gated discharge ports, each of 2.7 m diameter,

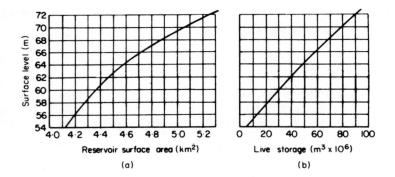

Figure 8.2 *Reservoir characteristics: (a) water level against area; (b) water level against storage*

whose centres are at elevation 54.0 and a free overflow spillway 72.5 m long with crest level at elevation 66.0. The discharge gates are open and the surface water level is at elevation 63.5 at time $t = 0$. The flood hydrograph of table 8.2, column (3), is forecast. What will the maximum reservoir level be and when will it occur?

(1) Assume the ports have a coefficient of discharge, $C_d = 0.8$, then $Q = 2(C_d A \sqrt{(2gH)})$ and at time 0, $Q = D$ (table 8.2, column (5)) $= 2(0.8 \times 5.7 \times \sqrt{186.5}) = 125 \text{ m}^3/\text{s}$. (Note $g = 9.81 \text{ m/s}^2$.) Insert this value on first line of column (5).

(2) Compute the elevation–storage curve of figure 8.2(b). Remember that live storage starts at 52.65, the invert level of the discharge ports, and amounts to $5.5 \times 10^6 \text{ m}^3$ by 54.0. The storage between 54.0 and 56.0 = mean area of reservoir between these levels $\times 2 \text{ m} = 8.33 \times 10^6 \text{ m}^3$. Successive increments, computed in this way, are plotted cumulatively with 52.65 as datum.

(3) Compute the elevation–D tabulation of table 8.1 below.

TABLE 8.1 *Elevation–discharge table*[a]

Elevation of water surface (m)	Head over 54.0 (m)	Discharge from gated ports (m³/s)	Head over 66.0 = H (m)	$H^{\frac{3}{2}}$	Spillway discharge (m³/s)	Total discharge (m³/s)
58.0	4.0	81.0	–	–	–	81
60.0	6.0	99.5	–	–	–	100
62.0	8.0	114.8	–	–	–	115
64.0	10.0	128.0	–	–	–	128
66.0	12.0	140.7	0	0	0	141
66.1	12.1	141.2	0.1	0.032	5.1	146
66.2	12.2	141.8	0.2	0.089	14.2	156
66.3	12.3	142.6	0.3	0.164	26	169
66.4	12.4	143.0	0.4	0.252	40	183
66.5	12.5	143.7	0.5	0.353	56	200
66.7	12.7	144.8	0.7	0.58	93	238
66.9	12.9	146.0	0.9	0.85	136	282
67.0	13.0	146.4	1.0	1.0	160	306
67.5	13.5	149.3	1.5	1.84	294	443
68.0	14.0	152.0	2.0	2.83	453	605

[a]Assume Q (spillway) $= CLH^{\frac{3}{2}}$ and use $C = 2.2 \text{ m}^{\frac{1}{2}}/\text{s}$.

(4) From figure 8.2(b) and table 8.1, the D-storage curve of figure 8.3 can now be drawn; that is, the central curve. The abscissa of figure 8.3 is graduated in 'storage units'. Each storage unit = routing period $\times 1 \text{ m}^3/\text{s}$. Since the forecast hydrograph of table 8.2, column (3) is given at 6-h intervals, it is convenient to make this the routing period. Then each storage unit = $6 \times 3600 \times 1 = 21.6 \times 10^3 \text{ m}^3 = \frac{1}{4} \text{ m}^3/\text{s}$ day. The use of these units is necessary to keep the dimensions of table 8.2, columns (4), (6) and (7) compatible.

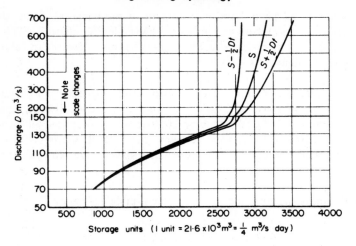

Figure 8.3   *Reservoir routing storage curves*

The $S \pm \frac{1}{2}Dt$ curves are now added on either side of the storage curve. Since the abscissa is in storage units, then $t = 1$ and the curves can be plotted without calculation; for example, at $D = 200, \frac{1}{2}Dt = 100$ and so two points can be set off 100 storage units on either side of the S-curve, and similarly for other points.

(5) The routing calculation can now be started in table 8.2. The figures in heavy type are known.
   To begin with, compute column (4) by averaging successive pairs of inflow values. Now find the value of this parameter corresponding to $D = 125 \, \text{m}^3/\text{s}$ from the $S - \frac{1}{2}Dt$ curve of figure 8.3. The value is 2095 and this figure is inserted in the first space of column (6).

(6) The figure in column (4) is added to column (6) and the total put in column (7) (for example, $2095 + 62 = 2157$). The left-hand side of equation 8.2 has now been evaluated. Find the column (7) value on the $S + \frac{1}{2}Dt$ curve and read off the corresponding value of D, entering it in column (5) (for example, from the $S + \frac{1}{2}Dt$ curve find $D = 122$ corresponding to 2157).

(7) Use this new value of D to find $S - \frac{1}{2}Dt$ again as in step (5). Note that this is directly obtainable without using the curve by subtracting the value of D from the column (7) value in the line above (for example, $2157 - 122 = 2035$). The column (4) figure is then added to the new column (6) to get a new column (7) figure (for example, $127 + 2035 = 2162$).

TABLE 8.2 *Reservoir routing computation*

(1)	(2)	(3)	(4)	(5)	(6)	(7)	(8)
Time (h)	Routing period	Inflow (m^3/s)	$\frac{(I_1 + I_2)t}{2}$	D (m^3/s)	$S - \frac{1}{2}Dt$ $(\frac{1}{4} m^3/s$ day$)$	$S + \frac{1}{2}Dt$ $(\frac{1}{4} m^3/s$ day$)$	Surface level (m)
0	1	50	62	125	2095	2157	63.4
6	2	75	127	122	2035	2162	63.0
12	3	180	265	122	2040	2305	63.0
18	4	350	400	127	2178	2578	63.8
24	5	450	485	136	2442	2927	65.1
30	6	520	512	200	2727	3239	66.5
36	7	505	475	425	2814	3289	67.4
42	8	445	402	460	2829	3231	67.5
48	9	360	325	416	2815	3140	67.35
54	10	290	270	347	2793	3063	67.15
60	11	250	230	288	2775	3005	66.95
66	12	210	192	242	2763	2955	66.7
72	13	175	157	208	2747	2904	66.55
78	14	140	125	190	2714	2839	66.45
84	15	110	97	165	2674	2771	66.25
90	16	85	75	144	2627	2702	66.05
96	17	65	60	140	2562	2622	66.0
102	18	55	52	138	2484	2536	65.3
108	19	50	47	134	2402	2449	64.7
114	20	45	42	132	2317	2359	64.3
120	21	40	39	129	2230	2269	64.0
126		38		127			63.7

Complete the table and plot the outflow hydrograph (figure 8.4). The peak outflow should fall on the recession limb of the inflow graph.

The time difference between the peaks of the inflow and discharge hydrographs is termed *reservoir lag* and the reduction in peak flows together with the spreading out of the recession curve is referred to as *attenuation*.

(8) The column (8) values of surface water level are derived from the values of discharge and levels in table 8.1. They are most conveniently found by plotting a graph and reading off the levels corresponding to the values of column (5).

The maximum water level in the case of this example is 67.5 m occurring about hour 40.

8.4 Routing in a river channel

The solution of the storage equation in this case is more complicated than for the simple reservoir, since wedge storage is involved. Storage is no longer a func-

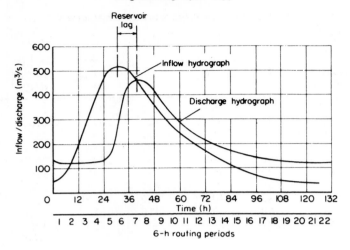

Figure 8.4 *Inflow and discharge hydrographs for the reservoir of example 8.1*

tion of discharge only, as was the case in example 8.1. McCarthy [1], in what has become known as the Muskingum method, proposed that storage should be expressed as a function of both inflow and discharge in the form

$$S = K[xI + (1 - x)D] \qquad (8.3)$$

where x = dimensionless constant for a certain river reach
        K = *storage constant* with dimensions of time that must be found from observed hydrographs of I and D at both stations.

The two constants can be found as follows. Let figure 8.5 represent the simultaneous inflow I and outflow D of a river reach. While $I > D$, water is entering

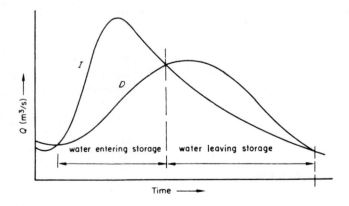

Figure 8.5 *Simultaneous inflow and outflow of a river reach*

storage in the reach and when $D > I$, water is leaving it. A difference diagram can now be drawn showing this (figure 8.6) and subsequently, a mass curve of storage (figure 8.7).

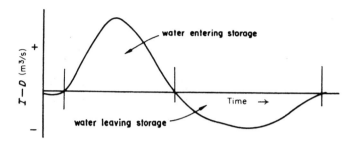

Figure 8.6 *Difference diagram for the river reach of figure 8.5*

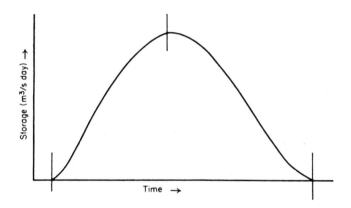

Figure 8.7 *Mass curve of storage for the river reach of figure 8.5*

Now assume a value of x, say $x = 0.1$, and compute the value of the expression $(0.1I + 0.9D)$ for various times and plot these against corresponding S values taken from figure 8.7. The resulting plot, known as a *storage loop* is shown in figure 8.8(a); clearly there is no linear relationship. Take further values of x (say 0.2, 0.3 etc.) until a linear relationship is established, as in figure 8.8(c) when the particular value of x may be adopted. K is now obtained by measuring the slope of the line.

Care about units is required. It is often helpful to work in somewhat unusual units, both to save computation and to keep numbers small. For example, storage S is conveniently expressed in m³/s day: such a unit is that quantity obtained from 1 m³/s flowing for 1 day = 86.4 × 10³ m³. If S is expressed in m³/s day and the ordinate of figure 8.8 is in m³/s then K is in days.

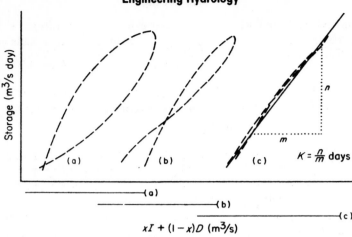

Figure 8.8 *River routing storage loops*

The following excerpt from Carter and Godfrey [2] concisely sums up the choice of values for x and K:

The factor x is chosen so that the indicated storage volume is the same whether the stage is rising or falling. For spillway discharges from a reservoir, x may be shown to be zero, because the reservoir stage, and hence the storage, are uniquely defined by the outflow; hence, the rate of inflow has a negligible influence on the storage in the reservoir at any time. For uniformly progressive flow, x equals 0.50, and both the inflow and the outflow are equal in weight. In this wave no change in shape occurs and the peak discharge remains unaffected. Thus, the value of x will range from 0 to 0.50 with a value of 0.25 as average for river reaches.

The factor K has the dimension of time and is the slope of the storage-weighted discharge relation, which in most flood problems approaches a straight line. Analysis of many flood waves indicates that the time required for the centre of mass of the flood wave to pass from the upstream end of the reach to the down-stream end is equal to the factor K. The time between peaks only approximates the factor K. Ordinarily, the value of K can be determined with much greater ease and certainty than that of x.

Having obtained values of K and x, the outflow D from the reach can be obtained, since by combining and simplifying the two equations

$$\frac{I_1 + I_2}{2} t - \frac{D_1 + D_2}{2} t = S_2 - S_1 \tag{8.1}$$

and

$$S_2 - S_1 = K[x(I_2 - I_1) + (1 - x)(D_2 - D_1)]$$

(the latter being equation 8.3 for a discrete time interval) the equation (8.4)

$$D_2 = C_0 I_2 + C_1 I_1 + C_2 D_1 \tag{8.4}$$

is obtained, where

$$C_0 = -\frac{Kx - 0.5t}{K - Kx + 0.5t}, \quad C_1 = \frac{Kx + 0.5t}{K - Kx + 0.5t}, \quad C_2 = \frac{K - Kx - 0.5t}{K - Kx + 0.5t} \quad (8.5)$$

where t = routing period, which should be taken as between $\frac{1}{3}$ and $\frac{1}{4}$ of the flood wave travel time through the reach (obtained from the inflow hydrograph).

A worked example illustrating the application of the method is set out below.

Example 8.2. Routing in a stream channel by the Muskingum method.
Part 1
Given the inflow and outflow hydrographs of figure 8.9, derive the constants x and K for the reach.

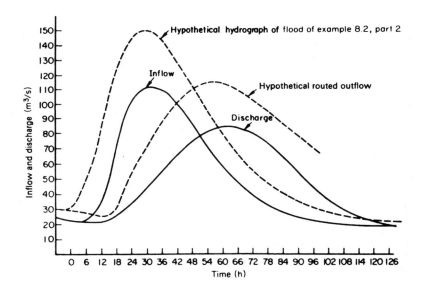

Figure 8.9 *Inflow and discharge hydrographs for a river reach*

The data are set out in tabular form in table 8.3. In columns 1 and 2, the given hydrographs are listed at a routing period interval, taken as 6 h. The storage units are taken here as ($\frac{1}{4}$ m³/s day) since the routing period is $\frac{1}{4}$ day. Colums (4), (5) and (6) are simply tabular statements of the processes illustrated in figures 8.6 and 8.7.

A value of x is then chosen, in the first instance 0.2, and the value inside the square brackets of equation 8.3 is then evaluated in columns (7), (8) and (9). Columns (6) and (9) are now plotted in figure 8.10 and produce the loop in the

TABLE 8.3 Storage loops calculations

(1)	(2)	(3)	(4)	(5)	(6)	(7)	(8)	(9)	(10)	(11)	(12)	(13)	(14)	(15)
						x = 0.2			x = 0.25			x = 0.3		
Hour	Inflow I (m³/s)	Outflow D (m³/s)	I−D (m³/s)	Mean storage (¼ m³/s day)	Cumulative storage (¼ m³/s day)	0.2I	0.8D	total	0.25I	0.75D	total	0.3I	0.7D	total
0	22	22	0	0	0	4	17	21	5	16	21	7	15	22
6	23	21	2	1	1	5	17	22	6	16	22	7	15	22
12	35	21	14	8	9	7	17	24	9	16	25	10	15	25
18	71	26	45	29	38	14	21	35	18	19	37	21	18	39
24	103	34	69	57	95	20	27	47	26	25	51	31	24	55
30	111	44	67	68	163	22	35	57	28	33	61	33	31	64
36	109	55	54	60	223	22	44	66	27	41	68	33	38	71
42	100	66	34	44	267	20	53	73	25	49	74	30	46	76
48	86	75	11	22	289	17	60	77	21	56	77	26	52	78
54	71	82	−11	0	289	14	66	80	18	61	79	21	57	78
60	59	85	−26	−18	271	12	68	80	15	64	79	18	59	77
66	47	84	−37	−31	240	9	67	76	12	63	75	14	59	73
72	39	80	−41	−39	201	8	64	72	10	60	70	11	56	67
78	32	73	−41	−41	160	6	58	64	8	55	63	10	51	61
84	28	64	−36	−38	122	6	51	57	7	48	55	8	45	53
90	24	54	−30	−33	89	5	43	48	6	40	46	7	38	45
96	22	44	−22	−26	63	4	35	39	5	33	38	7	31	38
102	21	36	−15	−18	45	4	29	33	5	27	32	6	25	31
108	20	30	−10	−12	33	4	24	28	5	22	27	6	21	27
114	19	25	−6	−8	25	4	20	24	5	19	24	6	17	23
120	19	22	−3	−4	21	4	18	22	5	16	21	6	15	21
126	18	19	−1	−2	19	4	15	19	4	14	18	5	13	18

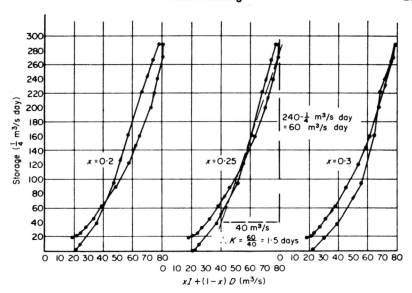

Figure 8.10 *Storage loops for the reach of example 8.2*

left-hand side of the figure.

A second value of $x = 0.25$ is now tried (columns (10), (11) and (12) refer) and the resulting plot is the central one of figure 8.10. A third value, $x = 0.3$, is also tabulated and plotted on the right of the figure. By inspection, the central value of $x = 0.25$ approximates a straight line most nearly, so this is chosen as the x value.

K is determined by measuring the slope of the median line as shown on the figure, and is found to be 1.5 days. This confirms the approximate peak-to-peak time of 33 h from figure 8.9. For this reach, therefore, use

$$x = 0.25 \text{ and } K = 1.5 \text{ days}$$

Part 2

Using the x and K values derived from the hydrographs, route the flood of table 8.4, column (2), through the reach and derive the outflow hydrograph.

First, compute C_0, C_1 and C_2 from equation 8.5. Use a routing period $t = 6$, $h = \frac{1}{4}$ day as before.

$$C_0 = -\frac{(1.5 \times 0.25) - (0.5 \times 0.25)}{1.5 - (1.5 \times 0.25) + (0.5 \times 0.25)} = -\frac{0.25}{1.25} = -0.2$$

TABLE 8.4   *Derivation of discharge*

(1)	(2)	(3)	(4)	(5)	(6)
Hours	I (m^3/s)	$-0.2\,I_2$ (m^3/s)	$0.4\,I_1$ (m^3/s)	$0.8\,D_1$ (m^3/s)	D_2 (m^3/s)
0	31				31.0[a]
6	50	−10.0	12.4	24.8	27.2
12	86	−17.2	20.0	21.8	24.6
18	123	−24.6	34.4	19.7	29.5
24	145	−29.0	49.2	23.6	43.8
30	150	−30.0	58.0	35.0	63.0
36	144	−28.8	60.0	50.4	81.6
42	128	−25.6	57.6	65.3	97.3
48	113	−22.6	51.2	77.8	106.4
54	95	−19.0	45.2	85.2	111.4
60	79	‍−15.8	38.0	89.1	111.3
66	65	−13.0	31.6	89.0	107.6
72	55	−11.0	26.0	86.1	101.1
78	46	−9.2	22.0	80.9	93.7
84	40	−8.0	18.4	74.9	85.3
90	35	−7.0	16.0	68.3	77.3
96	31	−6.2	14.0	61.8	69.6
102	27	−5.4	12.4	55.7	62.7
108	25	−5.0	10.8	50.2	56.0
114	24	−4.8	10.0	44.8	50.0
120	23	−4.6	9.6	40.0	45.0
126	22	−4.4	9.2	36.0	40.8

[a]Assumed value.

The similarly calculated values of $C_1 = 0.4$ and $C_2 = 0.8$ check that $-0.2 + 0.4 + 0.8 = 1.0$. From equation 8.4

$$D_2 = -0.2I_2 + 0.4I_1 + 0.8D_1$$

I_1, I_2 etc. are known from the hypothetical flood hydrograph, but D_1 is unknown. Assume a value for $D_1 = I_1 = 31\ m^3/s$. This will be very nearly correct since the river is at a low level and will be near to a steady state. Then the first equation to be solved is

$$D_2 = -0.2(50) + 0.4(31) + 0.8(31)$$

$$= -10.0 + 12.4 + 24.8 = 27.2$$

This value of D_2 becomes the D_1 for the next calculation and the values are tabulated as in table 8.4.

The outflow hydrograph is plotted as a dashed line to a little way beyond the peak in figure 8.9.

A disadvantage of the Muskingum method is that it does not include effects such as friction and diffusion, which can vary widely as a river reaches and then exceeds bank-full conditions. A method that does include these effects, known as the diffusion method, was devised by Hayami [3]. Cunge [4], however, showed that the solution to the Muskingum equations can be made to approximate to the solutions of the diffusion method provided the values of K and x are chosen appropriately. Essentially Cunge converted the method from one derived hydrologically to one based on hydraulic principles. He did this by writing the original Muskingum equations in finite difference form, identifying the magnitude of the error introduced and allowing this term to simulate the diffusion of the flood wave into storage areas along the reach.

In this modified method, now termed the Muskingum–Cunge method, the reach is divided into a series of sub-reaches, typically $L/10$ long, x is now derived for each sub-reach from the physical properties of the reach, i.e. average bed slope, mean channel width including the flood plain into which the flow diffuses, the average speed of the flood wave ω, and the average peak discharge. K is assumed to be $\Delta L/\omega$. The inflow hydrograph is then routed as before, through the sub-reaches in succession, with the values of K and x varying appropriately in each sub-reach.

For a more detailed description and further information about this method, the reader should consult references [5, 6].

8.5 Graphical routing methods

If equation 8.3 is written with $x = 0$, then

$$S = KD$$

and since, differentiating

$$\frac{dS}{dt} = K \frac{dD}{dt}$$

and

$$I - D = \frac{dS}{dt} \quad \text{(from section 8.2)}$$

then

$$\frac{I - D}{K} = \frac{dD}{dt}$$

This equation has been used [7] to provide a simple graphical method of routing, since dD/dt represents the slope of the outflow hydrograph and $(I - D)$ and K are measurable quantities in m^3/s and days. In figure 8.11, which is a plotted inflow hydrograph I, with discrete values I_1, I_2, I_3 etc. marked at intervals of time, the storage constant K is plotted horizontally from the position of each I value and a line drawn from the end of the K line to the previous discharge value D. Since this line represents dD/dt, the lower part can be used to denote the actual outflow hydrograph. Naturally, the smaller the time interval the more accurate the method will be, but there is no need to have the intervals equal.

K can be varied, if its variation is known, and reference to figure 8.10 will suggest K may well vary and could be obtained in a relationship with outflow from such a storage loop, giving a K against D curve as illustrated in figure 8.13(b).

The method can also be used in reverse, so that K at any time can be obtained from simultaneous hydrographs of I and D.

The foregoing description is all qualified by the initial statement that $x = 0$, so it applies to simple reservoir action only. It can be extended, however, to include positive x values since the effect of increasing x, with K constant, is to move the outflow graph bodily to the right so that the peak value no longer falls on the recession limb of the inflow graph, and also to increase the magnitude of the outflow peak.

If a succession of historic floods is analysed, the lag that is caused by x having a positive value can be determined. The lag due to this cause, T_x, is measured from the peak of the outflow graph to the same discharge on the recession limb of the inflow graph, as illustrated in figure 8.12 and a plot can be made connecting T_x with corresponding I (figure 8.13(b)).

Now the inflow graph of the reach with $x > 0$ is lagged as illustrated in figure 8.13, the amount of lag at each horizon being determined from the T_x against I curve, to give the dashed inflow graph, which is then routed by the graphical

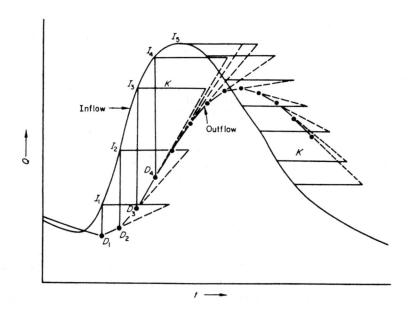

Figure 8.11 *Graphical routing method*

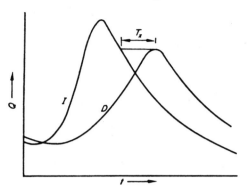

Figure 8.12 *Lag due to the constant x > 0*

method as though it were a simple reservoir inflow. It is often convenient to plot the curve of variation of K against D alongside the T_x against I curve so that both variations may be taken account of in the same plot. Fuller descriptions of this and similar routing techniques are available [8, 9].

8.6 Synthetic unitgraphs from flood routing

The principles of flood routing can now be used to derive unit hydrographs for a catchment where almost no rainfall or runoff records exist. The method is not entirely synthetic since at least one observation of a runoff hydrograph must be made.

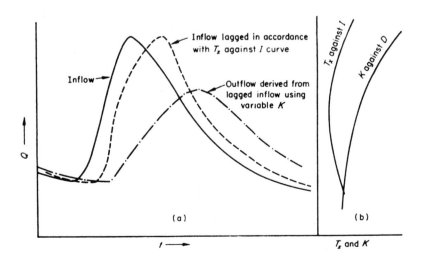

Figure 8.13 *Graphical routing with variable lag and K*

Consider a catchment as a series of sub-areas, each of which, under a sudden burst of rain, contributes inflow into the system of drainage channels, which have storage. The instantaneous unit hydrograph is therefore in two parts, the first representing the inflow of the rain, and the second the gradual withdrawal of the catchment storage. The dividing line between these two parts can be conveniently taken at the point of inflexion on the recession limb as shown in figure 8.14.

The assumption is now made that the catchment discharge Q and the storage S are directly proportional, so that

$$S = KQ \qquad (8.6)$$

(that is, equation 8.3 with $x = 0$, and Q used instead of D) and

$$I - Q = \frac{dS}{dt}$$

where I represents the inflow resulting from the instantaneous rain. Since $dS/dt = K\, dQ/dt$, by differentiating equation 8.6, then

$$K \frac{dQ}{dt} = I - Q$$

and using the condition $Q = 0$ when $t = 0$, the equation may be solved to

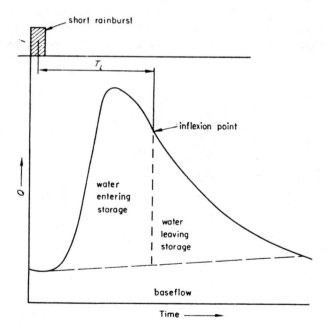

Figure 8.14 *Hydrograph from short rain approximate IUH*

$$Q = I(1 - e^{-t/K})$$

Since the inflow ceases at the inflexion point at time T (say), then the outflow at time t in terms of discharge Q_T at T is given as

$$Q_t = Q_T e^{-(t-T)/K} \tag{8.7}$$

The determination of the storage coefficient K must be made from an observed hydrograph on the catchment, as illustrated in figure 8.15, by taking two values unit time apart at the point of inflexion. The hydrograph should be of an isolated period of rain. It is not necessary that the magnitude of the rain be measured but it is necessary that it should be reasonably short, say 1 h only. Then

$$Q_1 = Q_T \text{ and from equation 8.7 } Q_2 = Q_T e^{-(t-T)/K}$$

the shaded area $A = \displaystyle\int_{t=T}^{t=T+1} Q_T e^{-(t-T)/K}$

Hence

$$A = \left[-KQ_T e^{-(t-T)/K} \right]_0^1$$
$$= K(Q_T - Q_T e^{-1/K})$$

so $$A = K(Q_1 - Q_2)$$

The second observation that must be made from the observed hydrograph is the catchment lag (T_L); that is, the maximum travel time through the catchment.

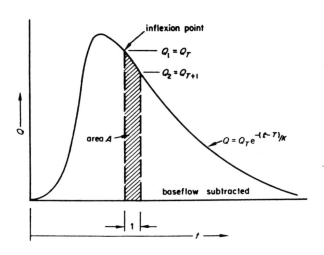

Figure 8.15 *Determination of K*

This may be taken as the time from the mass centre of the causative rain (hence the requirement that it should be short—so that no large error is introduced here) to the inflexion point on the recession limb.

The storage of the catchment is now thought of as a hypothetical reservoir, situated at the point of outflow; the inflow is expressed as the time–area graph of the catchment, where each sub-area is delineated so that all rain falling on it instantaneously has the same time of travel to the outflow point, as illustrated in figure 8.16. The time–area graph (I) now has instantaneous unit rain applied to it and is routed through the reservoir, in the manner of section 8.3, and the outflow (Q) derived. This outflow represents the IUH for the catchment and can be converted if required to the n-h unitgraph.

The method is basically due to Clark [10], though K is as derived by O'Kelly [11]. It is open to criticism in several respects and more advanced techniques [12-15] are now available, but it has the advantage of comparative simplicity. Its derivation is not dependent on an observed hydrograph of unit intensity.

Another advantage is that instead of deriving the IUH (and hence the n-h unitgraph) design rain can be applied directly to the time-area graph, with areal variation and in any desired quantity. This produces an instantaneous design-storm hydrograph, which can then be directly converted to a design-storm hydrograph of required intensity by averaging ordinates as discussed before.

A worked example of the method is given below.

Example 8.3. Given the catchment area of figure 8.16, of area 250 km², and the information derived from a short rain hydrograph that T_L = 8 h and K = 7.5 h, derive the 2-h unit hydrograph

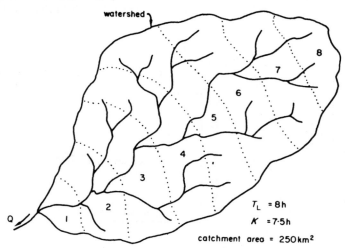

Figure 8.16 *Catchment with isochrones*

(1) Divide the catchment area into eight hourly divisions by *isochrones*, or lines of equal travel time. It will be assumed that all surface runoff falling in one of these divisions will arrive during a 1-h period at the gauging point.

(2) Measure by planimeter the area of each of the hourly areas. The areas of the figure are

Hour	1	2	3	4	5	6	7	8
Area (km^2)	10	23	39	43	42	40	35	18

(3) Draw the distribution graph of the runoff using the sub-areas as ordinates and 1-h intervals as abscissa. The result is figure 8.17—the *time-area graph* drawn in full lines.

(4) This time-area graph is now treated as the inflow I due to unit net rain of 1 cm on the catchment of a hypothetical reservoir, situated at the outlet, with storage equal to that of the catchment. Then

$$\frac{I_1 + I_2}{2} \cdot t - \frac{Q_1 + Q_2}{2} \cdot t = S_2 - S_1 \qquad \text{(from equation 8.1)}$$

and

$$S_1 = KQ_1 \quad \text{and} \quad S_2 = KQ_2 \qquad \text{(from equation 8.6)}$$

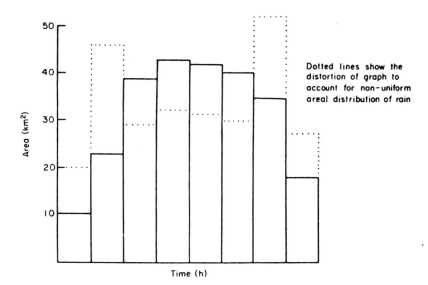

Dotted lines show the distortion of graph to account for non-uniform areal distribution of rain

Figure 8.17 *Sub-area distribution or time-area graph*

From these equations

$$Q_2 = m_0 I_2 + m_1 I_1 + m_2 Q_1$$

where

$$m_0 = \frac{0.5t}{K + 0.5t} \qquad m_1 = \frac{0.5t}{K + 0.5t} \qquad m_2 = \frac{K - 0.5t}{K + 0.5t}$$

and since a distribution graph is being used and $I_1 = I_2$, then

$$Q_2 = m'I + m_2 Q_1$$

where

$$m' = \frac{t}{K + 0.5t}$$

and in this case

$$m' = \frac{1}{7.5 + 0.5} = \frac{1}{8} = 0.125$$

and

$$m_2 = \frac{7.5 - 0.5}{7.5 + 0.5} = \frac{7}{8} = 0.875$$

Hence

$$Q_2 = 0.125 I + 0.875 Q_1$$

(5) Tabulate the data and compute Q_2 as in table 8.5. Q_2 is the required synthetic instantaneous unit hydrograph. Compute the conversion constant for column (3).

$$1 \text{ cm rain on } 1 \text{ km}^2 \text{ in } 1 \text{ h} = \frac{10^6 \times 10^{-2}}{3600} = 2.78 \text{ m}^3/\text{s}$$

(6) Plot the IUH and 2-h unitgraphs of table 8.5, columns (5) and (6) as figure 8.18.

To illustrate the ease of the method in accommodating areal variation in rainfall, suppose that a hypothetical rainfall is specified of 20 mm on sub-areas 1 and 2, 7.5 mm on 3, 4, 5 and 6, and 15 mm on areas 7 and 8, all falling in 1 h. The technique used then simply converts the time-area graph by these proportions, as shown by the dotted graph lines of figure 8.17 before the routing operation derives the IUH as before and then converts it to an n-h unitgraph by averaging each pair of ordinates at n-h spacing. In this latter case a degree of licence is being employed using the terms IUH and *unitgraph* since the rainfall is not uniform and catchment-wide as required by their definition.

TABLE 8.5 *IUH by routing*

(1)	(2)	(3)	(4)	(5)	(6)
Time (h)	Time area diagram (km²)	0.125I = 2.78 ×0.125 ×col. (2) (m³/s)	0.875 ×col. (5) (m³/s)	Q₂ = col. (3) + col. (4) = IUH (m³/s)	unitgraph (m³/s)
0	0	0	0	0	0
1	10	3.5	0	3.5	
2	23	8.0	3.1	11.1	5.5
3	39	13.5	9.7	23.2	13.4
4	43	14.9	20.3	35.2	23.1
5	42	14.6	30.8	45.4	34.3
6	40	13.9	39.6	53.5	44.3
7	35	12.1	46.8	58.9	52.2
8	18	6.2	51.4	57.6	55.5
9	0	0	50.5	50.5	54.7
10	0	0	44.1	44.1	50.8
11	0	0	39.6	39.6	45.1
12	0	0	34.6	34.6	39.3
13	0	0	30.2	30.2	34.9
14	0	0	26.4	26.4	30.5
15	0	0	etc.	etc.	etc.

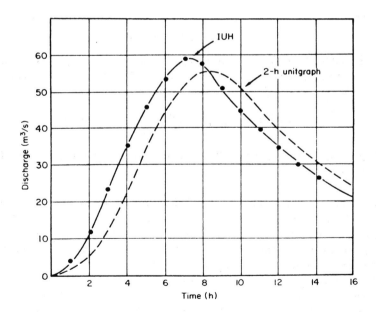

Figure 8.18 *Derived IUH and 2-h unitgraph*

References

1. McCARTHY, G. T. The unit hydrograph and flood routing. Unpublished paper presented at the Conference of the North Atlantic Division, Corps of Engineers, U.S. Army, New London, Connecticut, 24, June 1938. Printed by U.S. Engr. Office, Providence, Rhode Island
2. CARTER, R. W. and GODFREY, R. G. Storage and Flood Routing. *U.S. Geological Survey Water Supply Paper 1543-B*, 1960, p. 93
3. HAYAMI, S. On the propagation of flood waves. *Bulletin 1. Disaster Prevention Research Institute*, Kyoto University, Japan, 1951
4. CUNGE, J. A. On the subject of a flood propagation method. *J. Hydrol. Res.*, **7**, No. 2 (1969) 205–30
5. PRICE, R. K. Flood routing methods for British rivers. *Proc. Inst. Civ. Eng.*, **55** (1973) 913–30
6. Natural Environment Research Council. *Flood Studies Report*, Vol. III, NERC, 1975, Appendix 5.2
7. WILSON, W. T. A graphical flood routing method. *Trans. Am. Geophys. Union*, **21**, Part 3 (1941) 893
8. KOHLER, M. A. Mechanical analogs aid graphical flood routing. *J. Hydraulics Div.*, **84** (April 1958)
9. LAWLER, E. A. Flood routing. *Handbook of Applied Hydrology* (ed. Ven Te Chow), McGraw-Hill, New York, 1964, Section 25–II
10. CLARK, C. O. Storage and the unit hydrograph. *Trans Am. Soc. Civ. Eng.*, **110** (1945) 1419
11. O'KELLY, J. J. The employment of unit hydrographs to determine the flows of Irish arterial drainage channels. *Proc. Inst. Civ. Eng.*, Part III, (1955) 365
12. NASH, J. E. Determining runoff from rainfall. *Proc. Inst. Civ. Eng.*, **10** (1958) 163
13. NASH, J. E. Systematic determination of unit hydrograph parameters. *J. Geophys. Res.*, **64** (1959) 111
14. NASH, J. E. A unit hydrograph study, with particular reference to British catchments. *Proc. Inst. Civ. Eng.*, **17** (1960) 249
15. VEN TE CHOW. *Handbook of Applied Hydrology*, McGraw-Hill, New York, 1964, Section 14

Further reading

O'DONNELL, T. A. Direct three-parameter Muskingum procedure incorporating lateral inflow. *Hydrol. Sci. J.*, **30**, No. 4 (1985) 479
O'DONNELL, T., PEARSON, C. P. and WOODS, R. A. An improved three-parameter Muskingum routing procedure. *J. Hydraulic Engineering ASCE*, **114** (1987) 5
PRICE, R. K. A comparison of four numerical methods for flood routing. *J. Hydraulics Div. Am. Soc. Civ. Eng.*, **100**, HY7 (1974) 879–99
THOMAS, I. E. and WORMLEATON, P. R. Finite difference solution of the flood diffusion equation. *J. Hydrol.*, **12** (1971) 211–21

Problems

8.1 A catchment can be divided into ten sub-areas by isochrones in the manner shown in the table below, the catchment lag T_L being 10 h:

Hour	1	2	3	4	5	6	7	8	9	10
Area (km^2)	14	30	84	107	121	95	70	55	35	20

A single flood recording is available from which the storage coefficient K is found as 8 h. Derive the 2-h unit hydrograph for the catchment.

8.2 Tabulated below is the inflow I to a river reach where the storage constants are $K = 10$ h and $x = 0$:

Time (h)	I (m^3/s)	Time (h)	I (m^3/s)
0	28.3	40	90.6
5	26.9	45	70.8
10	24.1	50	53.8
15	62.3	55	42.5
20	133.1	60	34.0
25	172.7	65	28.3
30	152.9	70	24.1
35	121.8		

Find graphically the outflow peak in time and magnitude. What would be the effect of making $x > 0$? Assume outflow at hour 11 is 28.3 m^3/s and starting to rise.

8.3 A storm over the catchment shown in the figure generates simultaneously at A and B the hydrograph listed below:

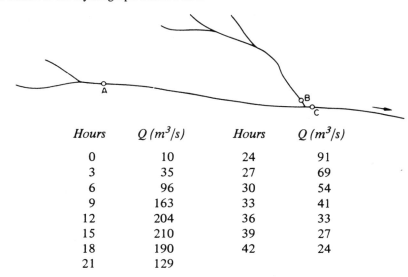

Hours	Q (m^3/s)	Hours	Q (m^3/s)
0	10	24	91
3	35	27	69
6	96	30	54
9	163	33	41
12	204	36	33
15	210	39	27
18	190	42	24
21	129		

Use the Muskingum stream flow routing technique to determine the combined maximum discharge at C. The travel time for the mass centre of the flood between A and C is 9 h and the factor $x = 0.33$. Any local inflow is neglected.

8.4 Define the instantaneous unit hydrograph of a catchment area, and describe how it can be used to derive the $n-h$ unitgraph.

A catchment area is 400 km² in total and is made up of the sub-areas bounded by the isochrones tabulated below:

Sub-area bounded by isochrone (h)	Area (km²)
1	15
2	30
3	50
4	75
5	80
6	60
7	45
8	25
9	20

From a short-storm hydrograph it is known that $T_L = 9$ h and the storage coefficient $K = 5.5$ h. Derive the 3-h unitgraph.

8.5 Listed below is the storm inflow hydrograph for a full reservoir that has an uncontrolled spillway for releasing flood waters:

3-h intervals	m³/s	3-h intervals	m³/s
0	1.5	12	54
1	156	13	45
2	255	14	40
3	212	15	34
4	184	16	28
5	158	17	23
6	136	18	17
7	116	19	11
8	99	20	8.5
9	85	21	5.5
10	74	22	3.0
11	62		

Determine the outflow hydrograph for the 48-h period after the start of the storm. Assume outflow is $1 \, m^3/s$ at time 0. The storage and outflow characteristics of the reservoir and spillway are tabulated below:

Height above spillway crest (m)	Storage (m³ × 10⁶)	Outflow (m³/s)	Height above spillway crest (m)	Storage (m³ × 10⁶)	Outflow (m³/s)
0.2	0.30	1.21	3.0	6.80	70.15
0.4	0.62	3.42	3.2	7.38	77.28
0.6	0.96	6.27	3.4	7.98	84.64
0.8	1.35	9.66	3.6	8.60	92.21
1.0	1.70	13.50	3.8	9.25	100.00
1.2	2.10	17.75	4.0	9.90	108.00
1.4	2.57	22.36	4.2	10.50	116.20
1.6	3.00	27.32	4.4	11.21	124.60
1.8	3.52	32.60	4.6	11.90	133.19
2.0	4.05	38.18	4.8	12.62	141.97
2.2	4.57	44.05	5.0	13.35	150.93
2.4	5.10	50.19	5.2	14.10	160.08
2.6	5.68	56.60	5.4	14.88	169.40
2.8	6.22	63.25			

8.6 Show by sketches how the Muskingum routing technique is based on at least one observation of a flood passing through a river reach. Compare the technique with reservoir routing to explain how and why the two techniques differ.

Determine the probable maximum discharge from a reach of a river where the inflow hydrograph is as listed below and the constants x and K of the routing equation are $x = 0.20$ and $K = 9 \, h$.

Hour	Q (m³/s)	Hour	Q (m³/s)
0	40	18	220
2	52	20	180
4	132	22	150
6	218	24	120
8	294	26	96
10	322	28	80
12	324	30	68
14	308	32	60
16	264	34	54

8.7 The following hydrograph was observed on a river as a result of an hour of uniform-intensity rain, baseflow having been estimated and subtracted:

Hour	Q (m³/s)	Hour	Q (m³/s)
0	0	8	603
1	40	9	582
2	120	10	540
3	265	11	467
4	405	12	385
5	515	13	306
6	580	14	232
7	607	15	167

The catchment contributing this flow was divided into sub-areas by isochrones, from which the following time–area correlation was derived:

Hour	1	2	3	4	5	6	7	8	9	10	11
Area (km²)	25	40	80	200	340	300	220	170	110	50	20

Given that $Q_2 = m'I + m_2 Q_1$ where $m' = t/(K + \frac{1}{2}t)$ and $m_2 = (K - \frac{1}{2}t)/(K + \frac{1}{2}t)$ derive the 1-h unit hydrograph for the catchment and determine Q_p and t_p.

Estimate the net rainfall that caused the original hydrograph.

8.8 The Muskingum routing equation $S = K(xI + (1 - x)D)$ applies to a reach of a river. Show how the constants x and K may be derived.

If the values for a particular reach are $K = 9$ h and $x = 0.30$, forecast the outflow hydrograph from the reach if this inflow hydrograph is as follows.

Hours	0	3	6	9	12	15	18	21	24	27	30	33	36	39	42	45	48
Inflow	6	5	17	48	81	102	105	103	95	64	45	34	27	20	16	13	12

9 Hydrological Forecasting

9.1 Introduction

In the previous chapters the various physical processes involved in the hydrologic cycle have been enumerated and examined in detail. Methods of evaluating each process have been suggested and often explained, and techniques discussed that can be used to provide quantitative answers to many questions.

The remaining problem that must now be tackled is how to use this knowledge to predict from existing data, however meagre it may be, what will happen in future. This is a fundamental problem of all engineering design, since the engineer designs and constructs work to provide for future needs, whether he be a structural engineer designing an office block, an electrical engineer designing power systems to meet future electrical demand, or a hydraulic engineer designing reservoirs to meet future demand for water.

There is one major difference in these three cases. The structural designer is working with homogeneous materials whose behaviour is known within narrow limits. His buildings will be used by people whose spacing, dimensions, weight and behaviour, *en masse*, can be predicted quite accurately. He has to cope with natural events only in the form of wind loads, which usually form a small proportion of the total load, and earthquakes. For both these eventualities there are codes of practice and recommendations available to him.

The electrical system designer has to extrapolate the rising demand curve of recent years, and to examine the trends of industry and personal habits, to decide how much capacity should be available in future years. While this is a complex and continuous task, it is almost completely immune from natural events other than disasters for which he cannot be expected to provide.

The hydraulic engineer on the other hand is dealing, in reservoir design, almost exclusively with natural events: in the incidence of precipitation, evaporation and so on. These events are usually *random* in nature and may have any or all the non-negative values. It is true that if the rainfall at a place is measured daily for a period of time, a knowledge about what is a probable daily rainfall

225

will be built up, but it will not, however long it goes on, lead to any limiting possible value of daily rainfall, other than intuitively.

The hydrologist is frequently asked what the maximum possible discharge of a particular river will be. There is simply no such value. The only answer that can be given is that from the data available, and making various assumptions, it would appear that a certain value will not be exceeded on average more than once in a specific number of years. On such estimates all hydrologic design must be performed, and this chapter deals with methods whereby some of the uncertainties can be removed or narrowed in range.

9.2 Flood formulae

9.2.1 Catchment area formulae. The particular random variable of river flood discharge has been of interest to engineers and hydrologists from the earliest days of hydrology and many formulae have been proposed to define the 'maximum flood' that could occur for a particular catchment. The formulae are empirical by nature, derived from observed floods on particular catchments and usually of the form

$$Q = CA^n$$

where Q = flood discharge in m^3/s (or ft.^{3}/s)
  A = catchment area in km^2 (or mile^2)
  n = an index usually between 0.5 and 1.25
  C = a coefficient depending on climate, catchment and units.

An early example of such a formula, due to Dickens, was developed in India

$$Q = 825a^{0.75}$$

with Q in ft.^{3}/s and a in square miles; but since the formula takes no account of soil moisture, rainfall, slope, altitude etc., it is clearly of very little value in general application. This is true of all such formulae although they are frequently used to obtain a quick first estimate of the order of 'maximum flood' that can be expected. For such purposes Morgan [1] proposed the formula for a catastrophic flood in Scotland and Wales of

$$Q = 3000M^{0.5}$$

where Q is in ft.^{3}/s and M is catchment area in square miles, and added the sophistication of a recurrence period T (in years) by quoting

$$\text{design flood} = \text{catastrophic flood} \times (T/500)^{\frac{1}{3}}$$

for cases where the adoption of the catastrophic flood was not justified by danger to human life or the safety of a dam. A similar formula of the same type, due to Fuller, has been widely used in the USA:

$$Q_{av} = CA^{0.8}$$

where A is catchment area in square miles
C is a coefficient often taken as 75
Q_{av} is average value of annual flood discharge in ft.^{3}/s.

The value of Q_{av} is then substituted in the formula

$$Q_m = Q_{av}(1 + 0.8 \log T)$$

where T is a return period in years and Q_m is the 'most probable' annual maximum flood.

9.2.2 The Rational Method. The introduction of rainfall into a formula might be expected to improve it, bearing in mind the type of relationship which exists between rainfall and runoff (figure 6.7 refers).

This kind of direct relationship of runoff to rainfall depths has been used in the past to determine flood discharges. Mulvaney [2] was the first to propose the idea in his work on Irish arterial drainage. It was also the basis of the Lloyd-Davis method of sewer designs [3] and the Bransby-Williams estimating method for floods in India [4]. Its use has persisted to the present because of its simplicity.

The formulae are all of the form

$$Q_p = CiA$$

where i = rainfall intensity in a time t
A = catchment area
t_c = time of concentration, i.e. the time taken for rain falling on the catchment farthest from the gauging station to arrive there
C = a dimensionless runoff coefficient, whose value depends on the catchment characteristics
Q_p = peak discharge due to the particular rainstorm, and assumed to occur after time t_c when the whole catchment area is contributing (it follows that the rain is assumed to be uniform over the catchment and to last at least for t_c).

Bransby-Williams gave a formula for the design rainfall duration D (in hours)

$$D = t_c = \frac{L}{d} \sqrt[5]{\left(\frac{a^2}{h}\right)}$$

where L = greatest distance from the edge of catchment to the outfall
d = is the diameter of a circle of area equal to catchment area
(L/d is, therefore, a dimensionless coefficient of circularity)
a = catchment area in square miles
h = the channel slope (as a percentage) along its greatest length
t_c = time of concentration in hours.

If i is in in./h, and A is in acres, then Q_p is in ft^3/s. Similarly if i is in mm/h and A in km^2 with Q_p in m^3/s, a correcting factor of 0.278 must be introduced to permit the same C values to be used, i.e.

$$Q_p \ (\text{ft}^3/\text{s}) = C \ i \ (\text{in./h}) \ A \ (\text{acres})$$

i.e. $$Q_p \ (\text{m}^3/\text{s}) = 0.278 \ C \ i(\text{mm/h}) \ A \ (\text{km}^2)$$

Many values are quoted for C, from 0.1 to 0.9 depending on the nature of the catchment, but as the runoff depends also on the intensity and duration of the storm, the catchment wetness etc., the method is of limited value unless data about a particular catchment are available. Then it may be used to interpolate or marginally extrapolate the data.

There is, however, one specialised use of the *Rational Method* where the value of C is sometimes taken as 1.0 i.e. completely impermeable ground. This is in the design of urban sewerage which is discussed in chapter 10.

A comparison of the *Rational Method* with those of the Flood Studies Report [5] indicated that the *Rational Method* typically yields results about twice the flood peak of the FSR for lowland catchments and much more as catchments become smaller and flatter.

9.2.3 A catchment parameter formula. A more sophisticated formula has been described by Farquharson *et al.* [6], based on estimates of maximum flow for 80 United Kingdom catchments made by the unit hydrograph method and subsequently related to catchment characteristics.

The estimated maximum flood (EMF) is given by

$$EMF = 0.835 \ \text{AREA}^{0.878} \ \text{RSMD}^{0.724} \ \text{SOIL}^{0.533} \ (1+\text{URBAN})^{1.308} \ \text{S1085}^{0.162}$$
(these variables are defined in section 9.4).

This formula takes account of catchment area, rainfall climate, slope of the drainage channel and the permeability of the surface soil, including any area which may be built up.

The great advantage of formulae is their simplicity and ease of use, but they are all of limited value since they apply to their derivative regions only, in most cases are derived from a limited series of recorded occurrences, and are 'envelope' estimates of indeterminate safety margins. They lack the precision obtainable from more refined methods and should be used only for preliminary estimates.

9.3 Frequency analysis

9.3.1 Series of events. The next approach is to use the methods of statistics to extend the available data and hence predict the likely frequency of occurrence of natural events. Given adequate records, statistical methods will show that floods of certain magnitudes may, on average, be expected annually, every 10 years, every 100 years and so on. It is important to realise that these extensions are only as valid as the data used. It may be queried whether *any* method of extrapolation to 100 years is worth a great deal when it is based on (say) 30 years of record. Still more does this apply to the '1000 year flood' and similar estimates.

Another point for emphasis is the non-cyclical nature of random events (see also section 9.6). The 100-year flood (that is, the flood that will occur *on average*, once in 100 years) may occur next year, or not for 200 years or may be exceeded several times in the next 100 years. The accuracy of estimation of the value of the (say) 100-year flood depends on how long the record is and, for flood flows, one is fortunate to have records longer than 30 years. Notwithstanding these warnings, frequency analysis can be of great value in the interpretation and assessment of events such as flood flows and the risks of their occurrence in specific time periods.

It is particularly important to define what is meant by an event. For example, if a river has been gauged every day for 10 years, there will be about 3650 observations. These are not independent random events since the flow on any one day is dependent to some extent on that of the day before, and so the observations do not comprise an *independent series*. The array of these observations is termed a *full series*.

Suppose the maximum event was extracted from the 10-year record in each year. These would constitute an independent series since it is highly unlikely that the maximum flow of one year is affected by that of a previous year. Even so, care is necessary, as can be seen from figure 9.1, where *water years*, measured between seasons of minimum flow, are marked, as well as calendar years. One

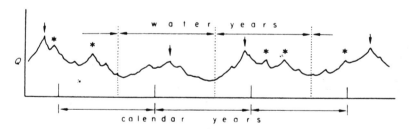

Figure 9.1 *Annual and partial duration series events*

calendar year might contain two water-year peaks, so it is necessary to specify that water years should be used in defining events. A selection like this is called an *annual series*. Such a series is open to the objection that some of the peaks are smaller events than secondary peaks (marked with an asterisk in figure 9.1) of other years. The objection can be overcome by listing a *partial duration series*, in which strict time segregation is no longer a condition and all peaks above some arbitrary value (say the lowest annual peak) are included provided that, in the judgement of the compiler, they are independent events, uninfluenced by preceding peak flows. Partial duration series events therefore allow the objection of subjective judgement and are not, strictly speaking, independent and random. The Flood Studies Report [3] gives a detailed account of a study of the partial duration, or peaks over a threshold (POT), series. It was concluded that the recommended POT model was a useful one for the estimation of the mean annual flood $(\overline{Q})$, particularly when the record was comparatively short, say from 3 to 10 years. A summary of the method is given later in section 9.3.6.

Which series is used depends on the purpose of the analysis. For information about fairly frequent events (for example, the size of a flood that might be expected during the construction period of a large dam—4 years say), then a partial series may be best, while for the design flood for the dam's spillway that should not be exceeded in the dam's lifetime (say 100 years) the true distribution series, or annual series, will be preferable. Actually, with very large floods there is a very small difference in *recurrence interval* between the two. Full series events, although not independent, are most valuable in design where quantity rather than peak values are required.

9.3.2 Probability of the N-year event.

The term *recurrence interval* (also called the *return period*), denoted by T_r, is the time that, on average, elapses between two events that equal or exceed a particular level. Putting it another way, the N-year event, the event that is expected to be equalled or exceeded, on average, every N years, has a recurrence interval, T_r of N years.

As mentioned previously there is no implication that the N-year event occurs cyclically. It does, however, have a probability of occurrence in any particular period under consideration.

Let the probability $P(X \leqslant x)$ represent the probability that x will not be equalled or exceeded in a certain period of time.

Then $P(X \leqslant x)_n$ will represent the probability that x will not be equalled or exceeded in n such periods.

For an independent series and from the multiple probability rule

$$P(X \leqslant x)_n = [P(X \leqslant x)]^n$$
$$= [1 - P(X \geqslant x)]^n$$

Therefore

$$P(X \geqslant x)_n = 1 - [1 - P(X \geqslant x)]^n$$

Now

$$T_r = \frac{1}{P(X \geqslant x)}$$

therefore

$$P(X \geqslant x)_n = 1 - \left[1 - \frac{1}{T_r}\right]^n$$

So, for example, the probability of $X \geqslant x$, where x is the value of a flood with a return period of 20 years, occurring in a particular 3-year period is

$$
\begin{aligned}
P(X \geqslant 20 \text{ yr flood})_3 &= 1 - [1 - \tfrac{1}{20}]^3 \\
&= 1 - [0.95]^3 \\
&= 1 - 0.857 \\
&= 0.143 \text{ or } 14.3 \text{ per cent}
\end{aligned}
$$

Table 9.1 shows the probability of the N-year flood occurring in a particular period.

For example, it can be seen from the table there is a 1 per cent chance of the 200-year flood occurring in the next 2 years and an 8 per cent chance that it will not occur for the next 500 years.

If the probability $P(X \leqslant x)_n$ is defined by a policy ruling, the value of n, the design period, can be found from

$$P(X \geqslant x)_n = 1 - \left(1 - \frac{1}{T_r}\right)^n$$

TABLE 9.1 *Percentage probability of the N-year flood occurring in a particular period[a]*

Number of years in period	N = Average return period T_r (years)							
	5	10	20	50	100	200	500	1000
1	20	10	5	2	1	0.5	0.2	0.1
2	36	19	10	4	2	1	0.4	0.2
3	49	27	14	6	3	1.5	0.6	0.3
5	67	41	23	10	5	2.5	1	0.5
10	89	65	40	18	10	5	2	1
20	99	88	64	33	18	10	4	2
30	99.9	96	78	45	26	14	6	3
60	—	99.8	95	70	43	26	11	6
100	—	—	99.4	87	63	39	18	10
200	—	—	—	98.2	87	63	33	18
500	—	—	—	—	99.3	92	63	39
1000	—	—	—	—	—	99.3	86	63

[a]Where no figure is inserted the percentage probability > 99.9.

$$1 - P(X \geqslant x)_n = \left(1 - \frac{1}{T_r}\right)^n = \left(\frac{T_r - 1}{T_r}\right)^n$$

$$\log (1 - P(X \geqslant x)_n) = n \log \left(\frac{T_r - 1}{T_r}\right)$$

Therefore

$$n = \frac{\log (1 - P(X \geqslant x)_n)}{\log \left(\dfrac{(T_r - 1)}{T_r}\right)}$$

Example 9.1.   How long may a cofferdam remain in a river, with an even chance of not being overtopped, if it is designed to be secure against a 10-year flood?

Here, the policy ruling is that there should be an even chance, so $P(X \geqslant x)_n = 0.50$ and $T_r = 10$ then

$$n = \frac{\log (1 - 0.5)}{\log \frac{9}{10}} = \frac{\log 0.5}{\log 0.9} = \frac{\bar{1}.699}{\bar{1}.954} = \frac{0.301}{0.046} = 6.5 \text{ years}$$

9.3.3 Determining the magnitude of the N-year event by plotting.   Having listed a series of events (for example, maximum floods) they are then each accorded a ranking m, starting with $m = 1$ for the highest value, $m = 2$ for the next highest and so on in descending order. The recurrence interval T_r can now be computed from one of a number of formulae, which have been reviewed by Cunnane [7].
   That most frequently used in the past is the Weibull formula [8]

$$T_r = \frac{n + 1}{m} \tag{9.1}$$

where m = event ranking and n = number of events, but there are objections to its use because of the bias it introduces to the largest events in a short series.
   Other formulae used are the Californian [9]

$$T_r = \frac{n}{m}$$

and Hazen's [10]

$$T_r = \frac{2n}{2m - 1}$$

about both of which there are reservations. One of the more satisfactory, due to Gringorten [11] is

$$T_r = \frac{(n + 0.12)}{(m - 0.44)} \tag{9.2}$$

For a single, simple compromise, Cunnane recommends

$$T_r = \frac{n + 0.2}{m - 0.4}$$

The probability P of an N-year event of return period T_r is

$$P = \frac{100}{T_r} \text{ per cent} \tag{9.3}$$

so that once the series has been ranked, its various events can be plotted on graphs connecting the variable Q and either T_r or P.

It is often assumed that such series are *normally distributed*, in which case the plotted points on *normal probability paper* would lie on a straight line. This seldom happens for flood series and shallow curves more often result, making extrapolations more difficult. To overcome this difficulty the variate Q is sometimes plotted logarithmically, which requires logarithmic-normal probability, or *log-normal* paper.

Table 9.2 is a listing of the annual maximum mean daily flows of the River Thames at Teddington Weir for the years 1883-1988. This is a true annual series with return periods and probabilities calculated from equations 9.2 and 9.3. Table 9.2 data can be plotted in a variety of ways and figures 9.2 to 9.5 illustrate the most common.

TABLE 9.2 *Maximum mean daily flows (Q_{max}) for water-years (ending 30 Sept.) 1883-1988, for the River Thames at Teddington[a]*

Water year	Q_{max} (m^3/s)	Rank m	Return period Tr (year)	Percentage probability P	Log Q_{max}
1883	511	11	10.1	10.0	2.708
1884	231	82	1.3	76.8	2.364
1885	230	85	1.3	79.7	2.362
1886	244	78	1.4	73.1	2.387
1887	284	65	1.6	60.8	2.453
1888	208	93	1.2	87.2	2.318
1889	237	80	1.3	75.0	2.375
1890	205	94	1.1	88.1	2.312
1891	171	100	1.1	93.8	2.233
1892	339	42	2.6	39.2	2.530
1893	300	58	1.8	54.2	2.477
1894	173	99	1.1	92.9	2.238
1895	1065	1	189.5	0.53	3.027
1896	202	96	1.1	90.0	2.305
1897	351	38	2.9	35.4	2.545
1898	171	101	1.1	94.8	2.233
1899	262	70	1.5	65.5	2.418

Water year	Q_{max} (m^3/s)	Rank m	Return period Tr (year)	Percentage probability P	Log Q_{max}
1900	533	7	16.2	6.2	2.727
1901	200	97	1.1	91.0	2.301
1902	162	102	1.0	95.7	2.210
1903	386	24	4.5	22.2	2.587
1904	517	10	11.1	9.0	2.713
1905	229	86	1.2	80.6	2.360
1906	249	76	1.4	71.2	2.396
1907	220	92	1.2	86.3	2.342
1908	376	28	3.9	26.0	2.575
1909	204	95	1.1	89.1	2.310
1910	231	83	1.3	77.8	2.364
1911	428	18	6.0	16.5	2.631
1912	367	36	3.0	33.5	2.565
1913	255	74	1.4	69.3	2.407
1914	256	73	1.5	68.4	2.408
1915	585	4	29.8	3.4	2.767
1916	373	31	3.5	28.8	2.572
1917	327	46	2.3	42.9	2.515
1918	351	39	2.8	36.3	2.545
1919	334	43	2.5	40.1	2.524
1920	251	75	1.4	70.3	2.400
1921	240	79	1.4	74.1	2.380
1922	198	98	1.1	91.9	2.297
1923	231	84	1.3	78.7	2.364
1924	298	60	1.8	56.1	2.474
1925	522	9	12.4	8.1	2.718
1926	370	32	3.4	29.7	2.568
1927	375	29	3.7	26.9	2.574
1928	526	8	14.0	7.1	2.721
1929	235	81	1.3	75.9	2.371
1930	552	6	19.1	5.2	2.742
1931	228	87	1.2	81.6	2.358
1932	274	66	1.6	61.8	2.438
1933	479	12	9.2	10.9	2.680
1934	95	106	1.0	99.5	1.978
1935	227	88	1.2	82.5	2.356
1936	478	13	8.5	11.8	2.679
1937	438	17	6.4	15.6	2.641
1938	247	77	1.4	72.1	2.393
1939	369	33	3.3	30.7	2.567
1940	410	19	5.7	17.5	2.613
1941	384	26	4.2	24.1	2.584
1942	298	61	1.8	57.1	2.474
1943	457	14	7.8	12.8	2.660
1944	115	105	1.0	98.5	2.061
1945	261	71	1.5	66.5	2.417
1946	257	72	1.5	67.4	2.410
1947	714	2	68.0	1.5	2.854

Water year	Q_{max} (m^3/s)	Rank m	Return period Tr (year)	Percentage probability P	Log Q_{max}
1948	227	89	1.2	83.5	2.356
1949	299	59	1.8	55.2	2.476
1950	324	49	2.2	45.8	2.511
1951	385	25	4.3	23.1	2.585
1952	377	27	4.0	25.0	2.576
1953	263	69	1.6	64.6	2.420
1954	222	91	1.2	85.3	2.346
1955	453	16	6.8	14.7	2.656
1956	316	52	2.1	48.6	2.500
1957	314	53	2.0	49.5	2.497
1958	317	51	2.1	47.6	2.501
1959	375	30	3.6	27.9	2.574
1960	308	56	1.9	52.4	2.489
1961	456	15	7.3	13.7	2.659
1962	344	41	2.6	38.2	2.537
1963	286	63	1.7	59.0	2.456
1964	369	34	3.2	31.6	2.567
1965	122	104	1.0	97.6	2.086
1966	324	48	2.2	44.8	2.511
1967	313	55	1.9	51.4	2.496
1968	600	3	41.5	2.4	2.778
1969	369	35	3.1	32.6	2.567
1970	224	90	1.2	84.4	2.350
1971	362	37	2.9	34.5	2.559
1972	330	45	2.4	42.0	2.519
1973	266	68	1.6	63.7	2.425
1974	396	22	4.9	20.3	2.598
1975	559	5	23.3	4.3	2.747
1976	152	103	1.0	96.6	2.182
1977	334	44	2.4	41.0	2.524
1978	326	47	2.3	43.9	2.513
1979	324	50	2.1	46.7	2.511
1980	393	23	4.7	21.3	2.594
1981	289	62	1.7	58.0	2.461
1982	314	54	2.0	50.5	2.497
1983	345	40	2.7	37.3	2.538
1984	286	64	1.7	59.9	2.456
1985	270	67	1.6	62.7	2.431
1986	408	20	5.4	18.4	2.611
1987	304	57	1.9	53.3	2.483
1988	402	21	5.2	19.4	2.604

[a]$Q_{av} = 329.7$ m^3/s, $n = 106$.
$\sigma = 133.8$.
T_r is calculated from
 equation 9.2.

$\Sigma X = \Sigma \log Q_{max} = 263.621$
$\Sigma X^2 = \Sigma (\log Q_{max})^2 = 658.478$
$X_{av} = 2.487$
$\sigma_x = 0.165$

(a) Q against T_r using linear co-ordinates (figure 9.2). Extrapolation of the curve to high values of T_r depends critically on the few highest points. Circled points are corresponding positions of equation 9.1 for two highest values.

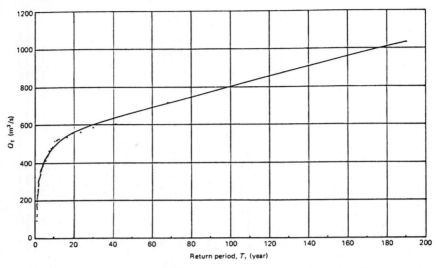

Figure 9.2 *Annual maximum mean daily flow of the River Thames at Teddington, 1883–1988*

(b) Q (linear) against T_r (logarithmic) (figure 9.3). This yields a straight line fitted to all but the lowest values. Although extrapolation is simpler, unless T_r follows a logarithmic law, extrapolation is not necessarily more accurate than for figure 9.2.

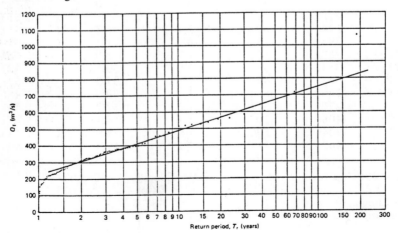

Figure 9.3 *Annual maximum mean daily flow of the River Thames at Teddington, 1883–1988 (semi-log)*

(c) Q (linear) against probability (per cent) (figure 9.4). As often happens, flood series points lie on a shallow curve on probability paper (where a normal distribution of probability is assumed).

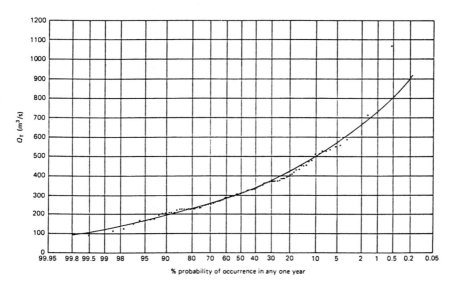

Figure 9.4 *Annual maximum mean daily flow of the River Thames at Teddington, 1883–1988 (normal probability)*

(d) Q (logarithmic) against probability (per cent) (figure 9.5). The curve of figure 9.4 is now transformed to a straight line.
   A variation of the approach in figure 9.4 is to assume that the logarithm of the variate Q is normally distributed, leading to the use of logarithmic-normal distribution (or *log-normal* paper)—first used by Whipple [12].

(e) Other investigators have proposed methods assuming other theoretical frequency distributions. Gumbel [13, 14, 15] used *extreme-value theory* (EV1) to show that in a series of extreme values $X_1, X_2 \ldots X_n$ where the samples are of equal size and X is an exponentially distributed variable (for example, the maximum discharge observed in a year's gauge readings), then the cumulative probability P' that any of the n values will be less than a particular value X (of return period T) approaches the value

$$P' = e^{-e^{-y}}$$

where e is the natural logarithm base

and

$$y = -\ln\left[-\ln\left(1 - \frac{1}{T}\right)\right]$$

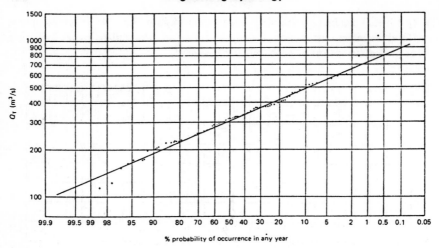

Figure 9.5 *Annual maximum mean daily flow of the River Thames at Teddington, 1883–1988 (log-normal)*

That is, P' is the probability of non-occurrence of an event X in T years, or

$$T = \frac{1}{1 - P'}$$

(Note that this argument refers to Gumbel's method. The reader should not confuse this with the normal usage of $T_r = 1/P$ where P = probability of occurrence.)

The event X, of return period T years, is now defined as Q_T, and

$$Q_T = Q_{av} + \sigma(0.78y - 0.45) \tag{9.4}$$

where Q_{av} = average of all values of 'annual flood' Q_{max}
        σ = standard deviation of the series.

Thus

$$\sigma = \sqrt{\left[\frac{n}{n-1} \left(\frac{\Sigma Q_{max}^2}{n} - Q_{av}^2 \right) \right]} \tag{9.5}$$

where n = number of years of record = number of Q_{max} values
       ΣQ_{max}^2 = sum of the squares of n values of Q_{max}.

Table 9.3 gives values of y as a function of T.

Powell [16] suggested that if plotting paper is prepared in which the horizontal lines are spaced linearly and the vertical lines' spacing is made proportional to y, then from equation 9.4 Q_T and T will plot as straight lines. This is the basis

TABLE 9.3 y as function of T

T	y	T	y
1.01	−1.53	100	4.60
1.58	0.00	200	5.30
2.0	0.37	300	5.70
5.0	1.50	400	5.99
10.0	2.25	500	6.21
20	2.97	1000	6.91
50	3.90	10,000	9.21

of Gumbel–Powell probability paper, used to plot the River Thames data in figure 9.6. The return period T has been computed, as before, from equation 9.2. The straight line on this figure has been drawn between the two points Q_{av} and Q_{200}. Q_{av}, from equation 9.4 occurs when $0.78y = 0.45$ or $y = 0.577$, which corresponds to $T = 2.33$ years. Equation 9.4 holds for large values of n, say

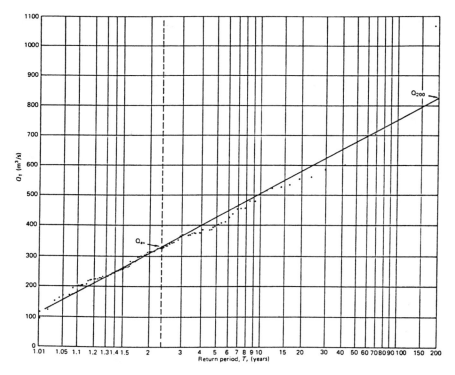

Figure 9.6 *Annual maximum mean daily flow of the River Thames at Teddington, 1883–1988 (Gumbel distribution)*

$n > 50$, when Q_{av} at 2.33 years is included on the line through the points. The other point Q_{200}, represents the '200-year flood' and is found by inserting the appropriate values in equation 9.4.

$$Q_{200} = Q_{av} + 133.8(0.78 \times 5.30 - 0.45) = 329.7 + 492.9 = 823 \text{ m}^3/\text{s}$$

and similarly

$$Q_{100} = Q_{av} + 133.8(0.78 \times 5.30 - 0.45) = 329.7 + 419.9 = 750 \text{ m}^3/\text{s}$$

The correspondence between the plotted data and Gumbel's theoretical line is demonstrated. Gumbel paper should not be used for partial series, which usually plot better on semi-log paper as used in figure 9.3.

From the plots presented in figures 9.2 to 9.6, it may seem there is little to choose between the particular plotting papers available. This is very often the case and investigators should use whichever distribution makes their particular job of fitting and extrapolation simplest and the line apparently of best fit.

The foregoing is a necessarily brief résumé of the methods of plotting flood events, in current use. For the underlying theories the reader should refer to original papers and more comprehensive treatments available [5, 7, 17, 18].

9.3.4 Determining the magnitude of the *N*-year event by calculation. Although the use of a normal probability distribution has been used above to plot events, and hence to extrapolate for rare values that may be used in design, values of particular probabilities can be calculated since a normal distribution curve is defined by only two parameters, the mean and standard deviation.

Accordingly, to determine the specific discharge associated with a particular probability of occurrence r in an annual series that is normally distributed, it is necessary to compute only

$$Q_r = Q_{av} + K\sigma$$

where σ = standard deviation from equation 9.5 and K is listed in table 9.4.

Example 9.1. Determine by calculation the mean daily discharge of the River Thames at Teddington with a 100-year return period, assuming the annual series is normally distributed.

From table 9.2: $Q_{av} = 329.7$ m^3/s and $\sigma = 133.8$ m^3/s. For $T_r = 100$ years, P is 1.0 per cent; and from table 9.4, $K = 2.33$. Therefore

$$Q_{100} = 329.7 + (2.33 \times 133.8)$$
$$= 641 \text{ m}^3/\text{s}$$

TABLE 9.4   *Values of a normal distribution*

Probability of exceedance (per cent)	K	Probability of exceedance (per cent)	K
0.1	3.09	50	0.00
0.5	2.58	55	−0.13
1.0	2.33	60	−0.25
2.5	1.96	65	−0.385
5	1.645	70	−0.52
10	1.28	75	−0.67
15	1.04	80	−0.84
20	0.84	85	−1.04
25	0.67	90	−1.28
30	0.52	95	−1.645
35	0.385	97.5	−1.96
40	0.25	99.0	−2.33
45	0.13	99.5	−2.58
50	0.00	99.9	−3.09

9.3.5 Log Pearson Type III distribution. It has already been mentioned that annual flood series are rarely normally distributed. A histogram of such series is usually *skewed*; that is, the mean value does not coincide with the *mode* (the value of the variate with largest frequency). Pearson [19] devised a measure of skewness based on (mean − mode)/σ and developed a family of curves to describe degrees of skewness. One of these, the Pearson Type III distribution, when used together with the logarithm of the variate Q is found to allow many annual flood series to plot as straight lines. This *log Pearson Type III* distribution has been adopted as a standard by several Federal agencies in the USA [20].

N-year events can be calculated in a similar manner to the normal distribution method, with this time the additional complication of using logarithms of the variate and a skew coefficient (G), given by

$$G = \left[\frac{n^2(\Sigma X^3) - 3n(\Sigma X)(\Sigma X^2) + 2(\Sigma X)^3}{n(n-1)(n-2)\sigma_x^3} \right] \tag{9.6}$$

where $X = \log Q$ and σ_x = standard deviation on n values of X.

Accordingly, to compute a particular return period flood Q from an annual series the following steps are required.

(a) Transform all (n) values of Q in the series to their logarithms (base 10):

$$X = \log Q$$

(b) Find the mean of all values of X:

$$X_{av} = \frac{\Sigma X}{n}$$

(c) Compute the standard deviation of n values of X

$$\sigma_x = \sqrt{\left[\frac{n}{n-1}\left(\frac{\Sigma X^2}{n} - X^2_{av} \right) \right]}$$

(d) Compute the skew of the X values, G, from equation 9.6.
(e) Calculate the discharge Q_r from ·

$$\log Q_r = X_{av} + K' \sigma_x \qquad (9.7)$$

where K' is selected from table 9.5 for the particular probability r and skew G.

TABLE 9.5   *Values of K' in the Pearson Type III distribution [21]*

Skew coefficient G	Probability of exceedance (per cent)						
	99	90	50	10	5	2	1
3.0	−0.667	−0.660	−0.396	1.180	2.003	3.152	4.051
2.5	−0.799	−0.771	0.360	1.250	2.012	3.048	3.845
2.0	−0.990	−0.895	−0.307	1.303	1.996	2.912	3.605
1.5	−1.256	−1.018	−0.240	1.333	1.951	2.743	3.330
1.2	−1.449	−1.086	−0.195	1.340	1.910	2.626	3.149
1.0	−1.588	−1.128	−0.164	1.340	1.877	2.542	3.023
0.9	−1.660	−1.147	−0.148	1.339	1.859	2.498	2.957
0.8	−1.733	−1.166	−0.132	1.336	1.839	2.453	2.891
0.7	−1.806	−1.183	−0.116	1.333	1.819	2.407	2.824
0.6	−1.880	1.200	−0.099	1.328	1.797	2.359	2.755
0.5	−1.955	−1.216	−0.083	1.323	1.774	2.311	2.686
0.4	−2.029	−1.231	−0.067	1.317	1.750	2.261	2.615
0.3	−2.104	−1.245	−0.050	1.309	1.726	2.211	2.544
0.2	−2.178	−1.258	−0.033	1.301	1.700	2.159	2.472
0.1	−2.253	1.270	0.017	1.292	1.673	2.107	2.400
0.0	−2.326	−1.282	0.000	1.282	1.645	2.054	2.326
0.1	−2.400	−1.292	0.017	1.270	1.616	2.000	2.253
−0.2	−2.472	−1.301	0.033	1.258	1.586	1.945	2.178
−0.3	−2.544	−1.309	0.050	1.245	1.555	1.890	2.104
−0.4	−2.615	−1.317	0.067	1.231	1.524	1.834	2.029
−0.5	−2.686	−1.323	0.083	1.216	1.491	1.777	1.955
−0.6	−2.755	−1.328	0.099	1.200	1.458	1.720	1.880
−0.7	−2.824	−1.333	0.116	1.183	1.423	1.663	1.806
−0.8	−2.891	−1.336	0.132	1.166	1.389	1.606	1.733
−0.9	−2.957	−1.339	0.148	1.147	1.353	1.549	1.660
−1.0	−3.023	−1.340	0.164	1.128	1.317	1.492	1.588
−1.2	−3.149	−1.340	0.195	1.086	1.243	1.379	1.449
−1.5	−3.330	−1.333	0.240	1.018	1.131	1.217	1.256
−2.0	−3.605	−1.303	0.307	0.895	0.949	0.980	0.990
−2.5	−3.845	−1.250	0.360	0.771	0.790	0.798	0.799
−3.0	−4.051	−1.180	0.396	0.660	0.665	0.666	0.667

Example 9.3. Determine by calculation, using a log Pearson Type III distribution, the annual maximum mean daily discharge of the River Thames at Teddington with a 100 year return period.

From table 9.2

ΣX = 263.621

ΣX^2 = 658.478

ΣX^3 = 1651.799

$(\Sigma X)^3$ = 18,320,613.4

X_{av} = 2.487

σ_x = 0.1648

Hence from equation 9.6

$$G = -0.066$$

From table 9.5 by interpolation at 1% probability (T_r = 100)

$$K' = 2.278 \quad \text{and hence} \quad K'\sigma_x = 0.375$$

and from equation 9.7

$$\log Q_{100} = 2.487 + 0.375 = 2.862$$

hence

$$Q_{100} = 728 \text{ m}^3/\text{s}$$

9.3.6 Estimation of the mean annual flood $\bar{Q}$ from a partial duration or peaks over threshold (POT) series [22]. When only a limited amount of data is available (say 3 to 10 years) a series of peaks over some arbitrary value may be used instead of an annual maxima series to estimate the value of $\bar{Q}$, the mean annual flood.

The procedure is as follows.

(1) Choose a level of threshold, q_0, such that on average between 3 and 5 peaks a year exceed it.

(2) Identify all peak flows above the threshold that are judged to be independent, using the rule that peaks should have a time separation of at least three times the average time to rise to a peak, where the average time is derived from five clean typical flood hydrographs in the record. Also, the minimum discharge in the trough between two peaks must be less than two thirds of the earlier peak value.

(3) Assume now that the number of exceedances per year can be treated as a Poisson variate whose parameter λ is given by

$$\lambda = M/N$$

and whose magnitude may be considered an exponential distribution whose parameter β is given by

$$\hat{\beta} = \bar{q} - q_0 = \sum_{i=1}^{M} (q_i - q_0)/M$$

(4) Q_T (the discharge with a return period of T years) can now be estimated from

$$Q_T = q_0 + \hat{\beta} \ln \lambda + \hat{\beta} \ln T$$

and

$$\bar{Q} = q_0 + \hat{\beta} \ln \lambda + 0.5772 \hat{\beta}$$

The theoretical background and justification for this is given in reference [5].

9.4 The FSR method of predicting $\bar{Q}$ and Q_T for an ungauged catchment

9.4.1 The general equation for $\bar{Q}$. This is a method for obtaining estimates of flood discharge for ungauged catchments through the use of catchment characteristics. The characteristic discharge adopted is the mean annual flood, $\bar{Q}$. After consideration of more than 500 catchments in Britain and Ireland and a detailed examination of the influence of catchment and meteorological characteristics on $\bar{Q}$, the recommended general equation for various hydrometric areas is

$$\bar{Q} = C [AREA^{0.94} \; STMFRQ^{0.27} \; SOIL^{1.23} \; RSMD^{1.03} \; S1085^{0.16} \; (1 + LAKE)^{-0.85}]$$

The notations are as follows. \hfill (9.8)

AREA: area in square kilometres.
STMFRQ: stream frequency (junctions/km^2) measured from OS 1:25 000
SOIL: soil index, which is a composite index determined from soil survey maps and is derived from the formula

$$SOIL = \frac{(0.15S_1 + 0.30S_2 + 0.40S_3 + 0.45S_4 + 0.5S_5)}{S_1 + S_2 + S_3 + S_4 + S_5}$$

where $S_1 \ldots S_5$ denote the proportions of the catchment covered by each of the soil classes 1–5. Soil class 1 has the highest infiltration capacity and hence lowest runoff potential, while soil class 5 has the lowest infiltration capacity and hence the highest runoff potential. SOIL has, therefore, a range of possible values between 0.15 and 0.5. (A full discussion of the soils classification may be found in FSR I 4.2.3, and maps in FSR V, figure 4.18.) A reduced sectionalised version of the maps for the British Isles is provided in appendix A, marked RP.

RSMD: the net 1-day rainfall of 5 years return period less s.m.d. This is found by obtaining 1-day M5 from 2-day M5 and ratio r (appendix A), tables 2.9 and 2.10, then table 2.8 and then subtracting mean s.m.d. (figure 4.8). Alternatively, an estimate may be obtained directly from figure 7.25, 7.26 or 7.27.

LAKE: the fraction of catchment draining through a lake or reservoir. Any lake or reservoir whose area is less than 1 per cent of the area contributing to the lake is ignored.

S1085: the stream slope (m/km) measured between two points situated at distances that are 10 per cent and 85 per cent of the stream length as measured from the catchment outlet along the longest stream length as marked on OS 1:25 000 maps.

C: a regional coefficient with values as indicated in figure 9.7.

For the Essex, Lee and Thames area only, a different equation is used

$$\bar{Q} = 0.373 \text{ AREA}^{0.70} \text{ STMFRO}^{0.52} (1 + \text{URBAN})^{2.5} \qquad (9.9)$$

where URBAN is the urban fraction of the catchment.

By using equation 9.8 (or 9.9) and the data provided in maps and figures, it is possible to predict $\bar{Q}$ for any catchment in the British Isles. Having obtained a value of $\bar{Q}$ for an ungauged catchment, an estimate of a flood Q_T, where T is the return period of a magnitude chosen for the design problem posed, is then required.

Figure 9.8 gives dimensionless region curves of $Q/\bar{Q}$ plotted against the reduced variate y from the Gumbel EV1 distribution. Return period T is also scaled. By identifying the particular region in which the catchment lies and using the appropriate region curve, the factor $Q/\bar{Q}$ is obtained for any particular return period and hence Q_T can be obtained. The regions are delineated and the values of coefficient C of equation 9.8 are given in figure 9.7.

Example 9.4. Estimate the flood frequency curve for the River Wyre at St. Michaels. This catchment is in Region 10.

AREA = 275 km²   STMFRO = 1.00   S1085 = 3.69
SOIL = 0.458   RSMD   = 44.5   LAKE = 0

The regional coefficient is 0.0213 so equation 9.8 becomes

$$\bar{Q} = 0.0213 \ [(275)^{0.94} \ (1.00)^{0.27} \ (0.458)^{1.23} \ (44.5)^{1.03} \ (3.69)^{0.16} \ ]$$
$$= 98.3 \text{ m}^3/\text{s}$$

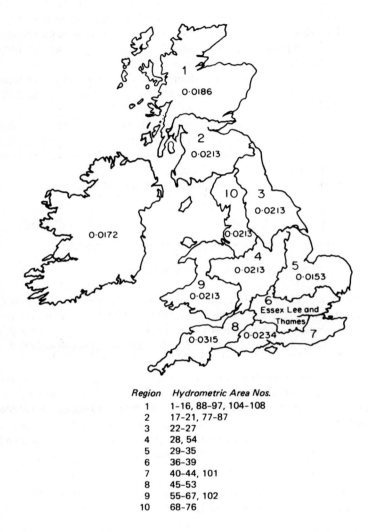

Region	Hydrometric Area Nos.
1	1–16, 88–97, 104–108
2	17–21, 77–87
3	22–27
4	28, 54
5	29–35
6	36–39
7	40–44, 101
8	45–53
9	55–67, 102
10	68–76

Figure 9.7 *Regions of the British Isles and corresponding values of C in equation 9.8 (derived from Flood Studies Report, Figure I.4.15)*

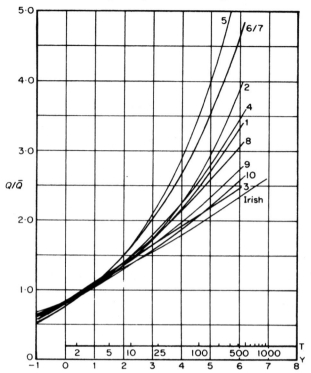

Figure 9.8 *Region curves showing average distribution of $Q/\bar{Q}$ for each region (derived from Flood Studies Report, Figure I.2.14)*

From figure 9.8 the $Q/\bar{Q}$ factors for Region 10 are listed and the flood frequency curve tabulated:

T_r (years)	$Q/\bar{Q}$	Q_T (m^3/s)
2	0.9	88
5	1.17	115
10	1.36	134
25	1.63	160
100	2.10	206
500	2.62	258

9.4.2 Small catchment equations for $\bar{Q}$. Since the publication of the Flood Studies Report, various authors have considered the possibility of predicting

Q for small catchments of less than 20 km^2 area with an equation of fewer terms than the six-term general equation (equation 9.8).

Many small catchments are not gauged and engineers designing for small bridges and culverts often have to resort to empirical equations. However, when catchments are small, the STMFRQ and S1085 terms are often difficult to quantify from standard maps.

Poots and Cochrane [23] studied 42 small catchments throughout the British Isles and concluded that the equation

$$\bar{Q} = 0.0136 \text{ AREA}^{0.866} \text{ RSMD}^{1.413} \text{ SOIL}^{1.521}$$

gave better results statistically than equation 9.8. Poots later concluded [24] that a slightly different version was a little better:

$$\bar{Q} = 0.015 \text{ AREA}^{0.882} \text{ RSMD}^{1.462} \text{ SOIL}^{1.904}$$

A similar exercise was carried out by the Institute of Hydrology on 47 catchments of similar size [25]. The equations resulting from regressing $\bar{Q}$ on catchment characteristics were for three terms and four terms respectively:

$$\bar{Q} = 0.00066 \text{ AREA}^{0.92} \text{ SAAR}^{1.22} \text{ SOIL}^{2.0} \tag{9.10}$$

and

$$\bar{Q} = 0.0288 \text{ AREA}^{0.90} \text{ RSMD}^{1.23} \text{ SOIL}^{1.77} \text{ STMFRQ}^{0.23} \tag{9.11}$$

In this case it was concluded that there was little difference between the prediction accuracy of equations 9.10 and 9.11 and the standard six-term equation 9.8, and that the former two should be used only if time was critical.

9.5 Synthetic data generation

One of the perennial problems of the hydrologist is insufficient data, whether it be rainfall or, more usually, discharge observations. If he wishes to predict flood flows of relatively large magnitude and hence large return periods, he may find that he has perhaps a decade or two of daily observations that represent discharges. Using an annual series such a record might yield 20 or 30 points from which it is clearly dubious to predict rare events of the order of the 100-year flood. By using only one or two of these measurements from each year's record, an enormous amount of information about the discharge characteristics of catchment is being left untouched. Is there not contained within this mass of routine observation a guide, not only to the catchment's response to rain, but to the incidence of the rain itself? By studying the fundamental nature of the variations in discharge, even at low flows, might it be possible to reproduce them, acting in their random and apparently uncoordinated ways and so automatically reproduce series exhibiting the variations that the natural data show?

The advent of powerful computers has provided answers to these questions by analysing and reproducing synthetic data in large quantities. While the quality of the synthetically generated data is mainly dependent on the original natural data, the method makes use of all the information available rather than the very small proportion of it in the form of extreme values.

In essence, the method of artificially generating time-series relies on using historical records as a sample of a total population, while conventional methods consider the records to be the total population. It follows that designs will be based on estimates of what might have happened instead of on what has happened.

Any time-series of observed values may contain a *trend* component, a *periodic* component and a *stochastic* component. The first two components are *deterministic* in nature (that is, they are not independent of the time at which a series starts nor of the length of the series) while the stochastic component is *stationary* (that is, the statistics of the sample do not differ from the statistics of the population, except as a result of sampling variability, and are time-independent).

If the trend and periodic components are removed from the series, a stationary stochastic component is left. This component will contain a random element and may or may not contain a correlation element. Series correlation describes how each term in a series is affected by what has gone before; for example, a wet summer may lead to higher autumn flows than average. Accordingly, the random element and correlation structure of the stochastic component must be isolated and quantified.

By now the time-series has been taken apart and its various parts examined. Each of the parts is now reproduced by mathematical simulation using randomly occurring numbers, Markov series, serial correlation coefficients etc., including the re-introduction of the periodicity and trend components. The 'model' thus created can now be used to generate synthetic data in whatever quantities are desired, and the series produced used to estimate particular N-year events as though the data had been observed.

9.6 The cyclical nature of hydrological phenomena

In all the foregoing sections of this book it has been tacitly assumed that the processes described and studied are based on non-changing physical conditions. For example, in the analysis of frequencies it is assumed that events that occurred in the last 50–100 years can be used to predict the probability of similar events occurring in future. From time to time this assumption is challenged but has rarely been disproved.

This is at least partly due to misconceptions about what periodicity implies, when associated with hydrological events. The implication of a periodicity in hydrological phenomena is that the likelihood of certain values of random events appears greater at certain times than at others. In other words there is a cyclical change in probabilities, rather than events. This does not rule out the possibility of maximum events (for example) occurring at times when their probabilities are least.

Brooks and Carruthers [26] describe the periodicity of annual rainfall in England, which seems to have a period of 51.7 years. They go on to show that the probability of a wet year was almost twice that of a dry year near the maximum of the 51.7 year cycle, whilst near the minimum it was less than half. This was true despite both very wet and dry years occurring at corresponding times of minimum probability.

It seems futile to employ statistical methods to derive probabilities of occurrence of certain events without recognising that certain processes, which are part-causative agents for these events, may be subject to cyclic probability themselves, thereby altering the derived probability to some degree. Although the opportunity for using such information may be rare, its presence should be investigated for any short-term probability analysis where the time-space is of the same order as the derivative data.

Cochrane, in an analysis of the hydrology of Lake Nyasa and its catchment [27] has demonstrated a correlation between the rate of change of sunspots and the 'free water' on Lake Nyasa. (The term 'free water' refers to the residual runoff from the catchment and storage on the lake, after losses have been deducted from rainfall.) In another paper [28] he quotes many references to support the case for cyclic behaviour in hydrologic phenomena, concerning work in many parts of the world. A dispassionate examination of the evidence leads the present writer to the conclusion that periodicity in hydrological phenomena exists, however imperfectly as yet we understand it.

References

1. MORGAN, H. D. Estimation of design floods in Scotland and Wales. *Symposium on River Flood Hydrology. Inst. Civ. Eng., London*, 1966, Paper 3
2. MULVANEY, T. J. On the use of self-registering rain and flood gauges. *Trans. Inst. Civ. Eng. Ireland*, 4 (No. 2) (1850) 1
3. LLOYD-DAVIES, D. E. The elimination of storm-water from sewerage systems. *Proc. Inst. Civ. Eng.*, 164 (1906) 41
4. BRANSBY-WILLIAMS, G. Flood discharge and dimensions of spillways in India. *Engineer*, 134 (1922) 32.1
5. Natural Environment Research Council. *Flood Studies Report*, Vol. I, NERC, 1975, pp. 185–213
6. FARQUHARSON, F. A. K., LOWING, M. J. and SUTCLIFFE, J. V. Some aspects of design flood estimation. *BNCOLD Symposium on Inspection, Operation and Improvement of Existing Dams, Newcastle University*, 1975 Paper 4.7
7. CUNNANE, C. Unbiased plotting positions—a review. *J. Hydrol.*, 37 (1978) 205–22
8. WEIBULL, W. A statistical theory of strength of materials. *Ing. Vet. Ak. Handl.*, 151, Stockholm, 1939
9. California Department of Public Works. Flow in California streams. *Bulletin 5, Calif. Dept. Pub. Wks.*, 1923
10. HAZEN, A. *Flood Flow*, John Wiley, New York, 1930

11. GRINGORTEN, I. I. A plotting rule for extreme probability paper. *J. Geophys. Res.*, **68** (1963) 813–14
12. WHIPPLE, G. C. The element of chance in sanitation. *J. Franklin Inst.*, **182** (1916) 37
13. GUMBEL, E. J. On the plotting of flood discharges. *Trans. Am. Geophys. Union*, **24**, Part 2 (1943) 699
14. GUMBEL, E. J. Statistical theory of extreme values and some practical applications. *National Bureau of Standards (U.S.) Appl. Math. Ser.*, **33** (February 1954)
15. Ref. 16, discussion by E. J. Gumbel, p. 438
16. POWELL, R. W. A simple method of estimating flood frequency. *Civ. Eng.*, **13** (1943) 105
17. VEN TE CHOW and YEVJEVICH, V. M. *Statistical and Probability Applied Hydrology* (ed. Ven Te Chow), McGraw-Hill, New York, 1964
18. DALRYMPLE, T. Flood Frequency Analysis. *U.S. Geological Survey Water Supply, Paper 1543-A*, Washington D.C., 1960
19. PEARSON, K. Tables for Statisticians and Biometricians, 3rd edition, Cambridge University Press, 1930
20. Guidelines for determining flood flow frequency. *Bulletin 17, Hydrology Communication Water Resources Council*, Washington D.C., June 1977
21. HARTER, H. L. A new table of percentage points of the Pearson Type III Distribution. *Technometrics* **11**, No. 1 (1969) 177–187
22. SUTCLIFFE, J. V. Methods of flood estimation: a guide to the Flood Studies Report. *Report No. 49, Institute of Hydrology*, Wallingford, United Kingdom, 1978
23. POOTS, A. D. and COCHRANE, S. R. Design flood estimation for bridges culverts and channel improvement works on small rural catchments. *Proc. Inst. Civ. Eng.*, **66**, TN 229 (1979) 663–6
24. POOTS, A. D. A flood prediction study for small rural catchments. Unpublished M.Sc. thesis, Queen's University, Belfast, 1979
25. Flood prediction for small catchments. *FSR Suppl. Report No. 6, Institute of Hydrology*, Wallingford, United Kingdom, 1978
26. BROOKS, C. E. P. and CARRUTHERS, N. *Handbook of Statistical methods in Meteorology*, H.M.S.O., London, 1953, p. 330
27. COCHRANE, N. J. Lake Nyasa and the River Shire, *Proc. Inst. Civ. Eng.*, **8** (1957) 363
28. COCHRANE, N. J. Possible non-random aspects of the availability of water for crops. *Conf. Civ. Eng. Problems Overseas, Inst. Civ. Eng.*, London, June 1964, Paper No. 3

Further reading

ALEXANDER, G. N. Some aspects of time series in hydrology. *J. Inst. Eng. Australia*, **26** (1954) 188–98
ANDERSON, R. L. Distribution of the serial correlation coefficient. *Ann. Math. Stat.*, **13** (1941) 1–13
BEARD, L. R. Simulation of daily streamflow. *Proc. Int. Hydrol. Symposium, Fort Collins, Colorado*, 6–8 September 1967, Vol. 1, pp. 624–32
FIERING, M. B. *Streamflow Synthesis*, Macmillan, London, 1967
HALL, M. J. and O'CONNELL, P. E. Time series analysis of mean daily river flows. *Water and Water Engineering*, **76** (1972) 125–33

HANNAN, E. J. *Time Series Analysis*, Methuen, London, 1960
KISIEL, C. C. Time series analysis of hydrologic data. *Advances in Hydroscience*, Vol. 5 (ed. Ven Te Chow), Academic Press, New York, 1969, pp. 1–119
LANGBEIN, W. B. Annual floods and the partial duration flood series. *Trans. Am. Geophys. Union*, **30** (December 1949) 879
MATALAS, N. C. Time series analysis. *Water Resources Res.*, 3 (1967) 817–29
MATALAS, N. C. Time series analysis. *Water Resources Res.*, **3** (1967) 817–29 *Resources Res.*, 3, (1967) 937–45
MORAN, P. A. P. *An Introduction to Probability Theory*, Clarendon Press, Oxford, 1968
National Research Council. *Estimating probabilities of extreme floods; Methods and recommended research*. Water Science and Technology Board, Commission on Eng. and Tech. Systems. National Academy Press, Washington D.C., 1989
NEWTON, D. W. Realistic assessment of maximum flood potentials *J. Hyd. Eng. ASCE*, **109**, No. 3 (1983) 905
O'DONNELL, T. Computer evaluation of catchment behaviour and parameters significant in flood hydrology. *Symposium on River Flood Hydrology, Inst. Civ. Eng.*, 1965
QUIMPO, R. G. Stochastic analysis of daily river flows. *J. Hydraulics Div., Am. Soc. Civ. Eng.*, **94** HY1 (1968) 43–57
Symposium on Hydrology of Spillway Design by the Task Force on Spillway Design Floods on the Committee on Hydrology. *Proc. Am. Soc. Civ. Eng.*, **90**, HY3 (May 1964)
WIESNER, C. J. Hydrometeorology and river flood estimation. *Proc. Inst. Civ. Eng.*, **27** (1964) 153
YEVJEVICH, V. M. and JENG, R. I. Properties of non-homogeneous hydrologic series. *Hydrology Paper No. 32, Colorado State University*, 1969
WANG, B. H. and REVELL, R. W. Conservatism of probable maximum flood estimates. *J. Hyd. Eng. ASCE*, **109**, No. 3 (1983) 400

Problems

9.1 A contractor plans to build a cofferdam in a river subject to annual flooding. Hydrological records over 30 years indicate a maximum flood flow of 7800 m³/s and a minimum of 2000 m³/s. The observed annual maxima plot as a straight line on semi-logarithmic paper where return period is plotted logarithmically.

The cofferdam will be in the river during four consecutive flood seasons and it is decided to build it sufficiently high to protect against the 20-year flood.

Evaluate (without plotting) the 20-year flood and determine the probability of its occurrence during the cofferdam's life.

9.2 The annual precipitation data for Edinburgh are given below for the years 1948–1963 inclusive:

Year	Precipitation (in.)	Year	Precipitation (in.)
1948	36.37	1956	28.17

1949	28.01	1957	25.68
1950	28.88	1958	29.51
1951	30.98	1959	18.04
1952	24.41	1960	24.38
1953	23.64	1961	25.29
1954	35.15	1962	25.84
1955	18.08	1963	30.24

(i) estimate the maximum annual rainfall that might be expected in a 20-year period and a 50-year period;

(ii) define the likelihood of the 20-year maximum being equalled or exceeded in the 9 years since 1963.

9.3 Discuss the methods in common use for plotting the frequency of flood discharge in rivers. List the separate steps to be taken to predict, for a particular river, the flood discharge with the probability of occurrence of 0.005 in any year. Assume 50 years of stage observations and a number of well-recorded discharge measurements with simultaneous slope observations.

9.4 The following table lists in order of magnitude the largest recorded mean daily discharges of a river with a drainage area of 12 560 km^2.

Year	Date	Discharge (m^3/s)	Year	Date	Discharge (m^3/s)
1948	29 May	2804	1922	19 May	1716
1948	22 May	2450	1925	20 May	1694
1933	10 June	2305	1924	13 May	1668
1928	26 May	2042	1917	9 June	1609
1932	14 May	2042	1916	19 June	1586
1933	4 June	2016	1912	21 May	1563
1917	17 June	1997	1918	5 May	1495
1947	8 May	1980	1918	10 June	1495
1917	30 May	1974	1929	24 May	1492
1921	20 May	1974	1943	29 May	1478
1927	8 June	1943	1922	26 May	1476
1928	9 May	1861	1919	23 May	1473
1927	17 May	1818	1936	10 April	1433
1917	15 May	1801	1936	5 May	1410
1938	19 April	1796	1923	26 May	1405
1936	15 May	1790	1927	28 April	1314
1922	6 June	1767	1939	4 May	1314
1932	21 May	1762	1934	25 April	1300
1912	20 May	1753	1945	6 May	1257
1938	28 May	1722	1935	24 May	1246

(continued overleaf)

Year	Date	Discharge (m^3/s)	Year	Date	Discharge (m^3/s)
1920	18 May	1235	1926	19 April	1017
1914	18 May	1195	1937	19 May	971
1931	7 May	1155	1944	16 May	969
1913	13 June	1119	1930	25 April	878
1940	12 May	1051	1941	13 May	818
1942	26 May	1051	1915	19 May	799
1946	6 May	1037			

The mean of annual floods is 1502 m³/s and the standard deviation of the annual series is 467 m³/s.

Compute return periods and probabilities for both partial and annual series. Plot the partial series data on semi-log plotting paper and the annual series on log-normal and Gumbel probability paper. Estimate from each the discharge for a flood with a probability of once in 200 years.

9.5 The annual rainfall in inches for Woodhead Reservoir for the period of record 1921–1960 is listed below:

Year	Rainfall	Year	Rainfall
1921	44.48	1941	46.79
1922	56.25	1942	43.38
1923	65.57	1943	45.87
1924	45.72	1944	59.00
1925	44.78	1945	42.74
1926	48.02	1946	55.39
1927	54.48	1947	41.04
1928	51.53	1948	45.03
1929	48.11	1949	44.11
1930	59.03	1950	52.15
1931	60.59	1951	55.43
1932	47.20	1952	48.87
1933	38.16	1953	43.26
1934	45.38	1954	63.13
1935	54.62	1955	36.46
1936	53.03	1956	56.57
1937	40.98	1957	51.79
1938	50.88	1958	51.22
1939	49.95	1959	39.58
1940	46.27	1960	60.66

The mean and standard deviation for the period are 49.69 in. and 7.08 in. respectively.

Arrange the data in ranking order. Compute return periods and probability. Plot the data on probability paper.

(a) What are the 50-year and 100-year annual rainfalls? How do these compare with prediction made using Gumbel's theory? What qualification needs to be made if using the latter results?

(b) What is the probability that the 20-year rainfall will be exceeded in a 10-year, a 20-year and a 40-year period?

(c) A certain waterworks plant is to be designed for a useful life of 50 years. It can tolerate an occasional rainfall year of 70 in. What is the probability that this amount may occur during the project's life?

9.6 A river is subject to annual flooding, and from the records of the observed annual maxima over 30 years, the data plot as a straight line on semi-log paper, with return period plotted on the logarithmic scale. The maximum recorded flood is 900 m³/s and the minimum is 150 m³/s. If 1000 m³/s is selected as a design discharge, what is the probability of its being exceeded during the next 20 years?

9.7 Below are the annual mean daily inflows (in $m^3 \times 10^6$) to a reservoir during 20 consecutive years:

| 7.31 | 6.90 | 6.64 | 5.08 | 5.37 | 5.75 | 7.30 | 7.22 | 6.48 | 5.20 |
| 6.38 | 5.51 | 5.65 | 5.82 | 5.81 | 6.30 | 6.53 | 6.12 | 6.06 | 6.07 |

The mean and standard deviation of the data are 6.175 and 0.68 ($\times 10^6$ m³) respectively

(a) Plot the data on Gumbel probability paper. Fit a straight line to the data and estimate the 100-year mean daily inflow. Compare plot with analytical value if $y = 4.60$ for $T = 100$ years.

(b) What is the probability of this 100 year value occurring in any period of 10 consecutive years?

9.8 Determine the magnitude of the 200-year flood from a catchment in the United Kingdom Region 2, given the following information:

catchment area   107 km²
stream frequency   1.32 junctions/km²
S1085   24.2 m/km
30 per cent of area is soil class 1, 35 per cent soil class 2, 20 per cent soil class 4 and 15 per cent soil class 5.
5 per cent of the catchment area drains through lakes
2-day M5 rainfall is 100 mm at catchment outlet

r = (60 minute M5/2-day M5) × 100 = 20 per cent

s.m.d. = 4 mm

What spillway capacity would you design for, with an earth dam, at this location—and why?

9.9 Annual flood discharges in a river for a 24-year period are listed below:

32.6	17.0	59.5
22.7	5.7	56.6
11.3	36.8	20.0
34.0	51.0	11.8
93.4	8.5	85.2
31.1	73.6	31.1
19.8	12.8	25.0
25.5	25.4	23.7

Rank the events in magnitude and compute their return periods and probabilities. Plot the data. Estimate the discharge of a 100-year flood.

What is the probability of such a flood occurring in the next 2 years?

9.10 The Flood Studies Report gives the following equation for a mean annual flood in the central region of England:

$$\bar{Q} = \frac{0.0213 \ AREA^{0.94} \ STMFRQ^{0.27} \ SOIL^{1.23} \ RSMD^{1.03} \ (1 + LAKE)^{-0.85}}{S1085^{0.16}}$$

(a) Discuss the significance of each term in the equation.

(b) Describe how you would use the equation to determine the flood with a T-year return period (Q_T).

(c) How would you amend Q_T for the design of a dam spillway? Indicate what value T might have in this case and whether you consider Q_T may be used directly.

(d) Indicate the relative merits of this method of flood determination and the unitgraph–rainfall approach. Give examples of situations where each method might be used.

9.11 Using the data listed in 9.4, determine by use of the log Pearson III distribution, the maximum annual mean daily discharge, with a return period of 50 years.

9.12 (a) Using the small catchment equation

$$\bar{Q} = 0.015 \ AREA^{0.882} \ RSMD^{1.462} \ SOIL^{1.904}$$

estimate the magnitude of 10 year, 100 year and 500 year floods from a catchment in Region 6 of

$$\begin{aligned} \text{AREA} &= 20 \text{ km}^2 \\ \text{RSMD} &= 50 \text{ mm} \\ \text{SOIL} &= 0.3 \end{aligned}$$

using figure 9.8.

(b) Compute an FSR synthetic unit hydrograph for the catchment if

$$\begin{aligned} \text{MSL} &= 4.5 \text{ km} \\ \text{S1085} &= 6.0 \text{ m/km} \end{aligned}$$

there is no urban development

If two successive 1-h periods of nett rain at an intensity of 3 mm/h separated by a 1-h dry period, fall on the catchment, what peak surface runoff would you expect? Ignore baseflow.

9.13 If $T_p = 46.6 \, (\text{MSL})^{0.14} \, (\text{S1085})^{-0.38} \, (1 + \text{URBAN})^{-1.99} \, (\text{RSMD})^{-0.4}$ use an FSR synthetic unit hydrograph to estimate the peak outflow of a 50 km² catchment where

$$\begin{aligned} \text{MSL} &= 8 \text{ km} & & \text{no urban} \\ \text{S1085} &= 5.0 \text{ m/km} & & \text{development} \\ \text{RSMD} &= 75 \text{ mm} \end{aligned}$$

when the catchment is subjected to a 7 h net rainfall of following intensity

$$\begin{aligned} \text{Hour 1 and 7} & \quad 0.2 \text{ cm/h} \\ \text{2 and 6} & \quad 0.4 \text{ cm/h} \\ \text{3 and 5} & \quad 0.8 \text{ cm/h} \\ \text{4} & \quad 1.5 \text{ cm/h} \end{aligned}$$

and baseflow rises from 2 m³/s at the beginning of rainfall to 5.5 m³/s at its end, thereafter declining at the same rate. Assume response runoff starts simultaneously with rainfall.

9.14 The maximum annual discharges for a river are listed below for thirty-three years of observation. The mean and standard deviation for the period are 131.5 m³/s and 83.0 m³/s respectively. Rank the data, compute probability and plot the data on log-normal probability or semi-log paper.

Year	Q_{peak} (m^3/s)	Year	Q_{peak} (m^3/s)
1928	104	1944	57
1929	48	1945	229
1930	95	1946	129
1931	46	1947	153
1932	101	1948	137
1933	311	1949	54
1934	72	1950	102
1935	127	1951	137
1936	203	1952	196
1937	50	1953	114
1938	89	1954	161
1939	181	1955	294
1940	88	1956	360
1941	81	1957	20
1942	160	1958	32
1943	143	1959	226
		1960	40

(a) Plot the line of best fit to the data and predict the 100-year and 400-year flood discharges.

(b) Given Gumbel's equation $Q_T = Q_{av} + \sigma(0.78y - 0.45)$ and that $y = 4.60$ for $T = 100$ year and $y = 5.99$ for $T = 400$ year, compare the results from Gumbel's theory with your plot.

(c) What probability is there of a flow equalling or exceeding 300 m³/s in the next five years?

9.15 The following table lists annual peak floods for a river gauging station. Order the events, compute probabilities and plot the results on log-normal probability paper.

Use $T_r = \dfrac{n + 0.2}{m - 0.4}$.

Year	Flood flow (m^3/s)	Year	Flood flow (m^3/s)
1945	1150	1957	445
1946	780	1958	2600
1947	530	1959	300
1948	1200	1960	870
1949	3300	1961	2100
1950	1100	1962	2000
1951	720	1963	700
1952	900	1964	580
1953	600	1965	3010
1954	420	1966	1100
1955	1300	1967	850
1956	1800	1968	800

What is the probability of a flood equal to or greater than 2900 m^2/s occurring in any year, and in any period of four consecutive years?

9.16 covers material from Chapters 1–9, and is included as an example of the multiple-choice type.

9.16 A list of possible answers, each identified by a number, is supplied with each of the questions. The numbers identifying the correct answers should be selected and written in the boxes provided beside the individual questions. The number of boxes beside the individual questions clearly indicates the number of answers required.

1. wind speed	22. Log Pearson 3 distribution
2. isohyets	23. Gumbel distribution
3. daily stage records	24. normal probability
4. peak discharge	25. flow measuring structure
5. flow duration curve	26. aquifer permeability
6. recording rain gauge	27. hydrograph base length
7. ambient temperature	28. drawdown
8. routing techniques	29. return period
9. psychrometer	30. time to peak flow
10. current meter	31. mass curve
11. cloud cover	32. M2
12. standard rain gauge	33. draw-off rate
13. dilution gauging	34. consumptive use
14. recession curve	35. S-curve
15. relative humidity	36. reservoir capacity
16. phreatic surface	37. compensation flow
17. velocity–area method	38. unit hydrograph
18. M5	39. mean annual flood
19. sub-surface geology	40. Thiessen polygons
20. solar radiation	41. areal reduction factor
21. vegetation	42. M1

Which of the above

☐ allows rainfall intensity to be estimated?

☐☐☐ are used to estimate areal rainfall?

☐☐☐☐ are used to establish a rating curve for a river?

☐☐☐☐☐ are used in calculating daily evaporation?

☐ is reference frequency for rainfall in the FSR methods?

☐☐ affects the quantity of baseflow?

☐☐☐ are the parameters for drawing a synthetic unit hydrograph?

☐ is an expression of how baseflow alters with time?

☐☐☐☐ are used in the capacity calculations for reservoirs?

☐☐ would be used in reservoir spillway design?

☐ is the method where 'equilibrium flow' is used as a check on unit hydrograph integrity?

☐☐☐ are used in hydro-electricity capacity design?

☐☐ are used with borehole pumping tests to establish reliable yield?

☐☐☐ are used to plot independent flood events and hence to extrapolate to rare frequencies?

☐ may be preferred for the gauging of minor streams?

☐☐ is important in the design of irrigation schemes?

☐ reduces evaporation?

☐ is used to change unitgraph ordinates when the duration of causative rain changes?

10 Urban Hydrology

10.1 Introduction

In the estimation of runoff from natural catchments, the determination of T_p, time to peak, and *SPR*, standard percentage runoff, contain an URBAN term. Similarly in chapter 9, the equation for $\bar{Q}$, the mean annual maximum flood for Region 6, departs from the general equation and includes an URBAN term. URBAN, it may be recalled, is the fraction of catchment in urban development. The reason for this special term's inclusion is the need to take account of the impermeable areas of buildings' roofs, roads, pavements, car parks, etc. in a built-up environment where there is negligible infiltration and runoff is accelerated by drainage systems of gulleys, pipes and sewers. Obviously the time to peak of a hydrograph and the percentage runoff will be decreased and increased respectively compared with a natural catchment of permeable soils and vegetation.

10.2 The use of the Rational Method

It is in the specialised hydrology of urban drainage systems that a particular use is found for the Rational Method of flood estimation. It was pointed out in the discussion of this method in chapter 9 that the difficulty of evaluating the coefficient C made the method of limited value, but if $C = 1.0$, i.e. complete impermeability, then the logic of the contributing area times the rainfall becoming runoff, allows the sewer designer to make reasonable estimates of pipe sizes. Sewer design is based on the system coping with high-intensity rainstorms of some pre-selected frequency and duration, and for drainage areas like housing estates and villages the Rational Method of Lloyd-Davies [1], subsequently modified, forms the basis of the TRRL Road Note 35 [2].

The recommended method involves the following procedures:

(a) A plan of the proposed sewer network is prepared, as shown typically in figure 10.1 and the component pipes numbered according to a decimal

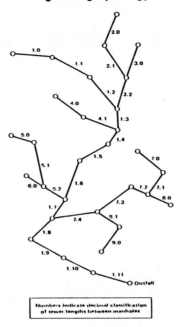

Figure 10.1 *Key plan of sewer system (reproduced from TRRL Road Note 35 (1976))*

system. The longest sewer (measured from the outfall) is numbered 1 and individual pipe lengths numbered successively 1.0, 1.1, 1.2, etc., starting at the upstream end. The first branch sewer which joins 1 is numbered 2 and the longest sewer contributing to the branch has its component parts numbered 2.0, 2.1, 2.2, etc. This process is repeated until every pipe in the system is numbered.

(b) A table is then prepared, of which table 10.1 is a typical example. The data of columns 1, 2, 3, 4, 9, 10 and 11 are then entered in it. All of this information should be available from a site survey and building proposals and plans. Notice that it is only the directly-contributing areas which are used. Permeable areas of gardens, woods, etc. are not included.

(c) Column 12 is then completed by adding in branch contributions at junctions as the table is completed. Column 12 gives the total surface area contributing to a particular length of sewer.

(d) It is now necessary to find the time of concentration, i.e. the time it takes for rain falling on the small area of impermeable surface that contributes to pipe 1.0 to arrive at the manhole at its downstream end. To do this it is necessary to assume a particular pipe diameter and material for 1.0. An entry time must be chosen and 2 minutes is recommended for normal urban areas, increasing to 4 minutes for very large, flattish paved areas. Knowing the pipe diameter and gradient, the full-bore velocity in the

TABLE 10.1 Rational formula design sheet[a],[b]

1 Pipe length number	2 Difference in level (m)	3 Length (m)	4 Gradient (1 in)	5 Velocity (m/s)	6 Time of flow (min.)	7 Time of concentration (min.)	8 Rate of rainfall (mm/h)	Impermeable area ha				13 Rate of flow (l/s)	14 Pipe diameter (mm)	15 Remarks
								9 Roads	10 Buildings, yards etc.	11 Total	12 Cumulative			
										(9 + 10)				
1.0	1.10	63.1	57	1.33	0.79	2.79	67.9	0.089	0.053	0.142	0.142	26.8	150	
1.1	1.12	66.1	59	1.70	0.65	3.44	62.5	0.077	0.109	0.186	0.328	56.9	225	
1.2	0.73	84.7	116	1.46	0.97	4.41	57.4	0.081	0	0.081	0.409	65.2	300	
2.0	1.40	44.8	32	2.32	0.32	2.32	72.5	0.113	0.081	0.194	0.194	39.1	225	
2.1	0.61	49.1	80	1.77	0.46	2.78	67.9	0.045	0.105	0.150	0.344	64.9	300	
3.0	0.98	48.5	49	1.43	0.56	2.56	70.5	0	0.129	0.129	0.129	25.2	150	
2.2	1.65	54.3	33	2.74	0.33	3.11	65.5	0.101	0.073	0.174	0.647	117.7	300	Sewer 3 added
1.3	1.22	27.7	23	3.29	0.14	4.55	56.9	0.121	0.235	0.356	1.412	223.2	300	Sewers 2 and 3 added
4.0	0.88	54.9	62	1.66	0.55	2.55	70.5	0.093	0.093	0.186	0.186	36.4	225	
4.1	0.58	45.7	79	1.48	0.52	3.07	66.3	0.069	0.040	0.109	0.295	54.2	225	
1.4	0.52	22.9	43	2.77	0.14	4.69	55.8	0.069	0	0.069	1.776	275.3	375	Sewer 4 added

[a] Reproduced from TRRL Road Note 35 (1976).
[b] Storm frequency one in 1 year. Roughness coefficient 0.6 mm. Time of entry 2 minutes.

pipe may be found from published tables [3] using the Colebrook–White formula and an appropriate k value in that formula, for the pipe material. This value is entered in column 5 and with the pipe length of column 3, the time of flow may be calculated and entered in column 6.

(e) The rate of rainfall corresponding to the time of concentration in column 7 (time of entry and time of flow in the case of the first pipe) is then determined from Meteorological Office tables. These are based on the Meteorological section of the Flood Studies Report and may be obtained for any National Grid reference point (approximately in the centre of the drainage area) from the Meteorological Office, London Road, Bracknell, Berkshire. Note that it is necessary to choose a design rain frequency. Suitable frequencies may be between one year for modern separate surface water systems and 100 years for old combined systems with basement developments. Table 10.2 is an abbreviated typical example.

TABLE 10.2 *Rates of rainfall in mm/h for a range of duration and return period for a specified location in the UK (in Southern England)[a]*

Duration (min.)	Return period (years)			
	1	2	10	100
2.0	75.6	93.4	138.3	213
3.0	66.3	82.3	123.4	192
5.0	54.3	67.1	102.5	163
8.0	43.4	53.4	82.9	135
12.0	35.0	42.8	66.9	111
20.0	25.9	31.4	49.3	83

[a]Derived from table 7 of *TRRL Road Note 35* (1976).

(f) Using the Rational formula

$$Q\,(\mathrm{m^3/s}) = 0.278 C_i(\mathrm{mm/h}) \times A\,(\mathrm{km^2}) \quad \text{or}$$

$$Q(\mathrm{l/s}) = 2.78 C_i(\mathrm{mm/h}) \times A\,(\text{hectares})$$

and taking C as 1.0, the discharge requiring conveyance in the pipe is determined and compared with the available capacity of the assumed pipe in column 6. If the latter is insufficient, the next larger pipe diameter is assumed and the processes of steps (d) and (e) repeated.

(g) When pipe 1.0 has been chosen and found acceptable, its discharge is added to the inflow to the next pipe automatically through the cumulative-area-contributing column (column 12). The method of working through the network is best seen by inspection of table 10.1.

There are many other recommendations and rules for the optimum design of such completely separate surface-water systems and methods of incorporating additional permeable-area drainage, for which reference [2] should be consulted.

Although the procedure of the TRRL Rational Method has been described to illustrate the principles involved, the Modified Rational Method of the Wallingford Procedure [4] has refined and improved the method. The modifications include, amongst others, a value of the coefficient C derived from

$$C = C_v \times C_r$$

where C_v is the volumetric runoff coefficient and C_r is a routing coefficient. C_v varies from 0.6 to 0.9, depending on the nature of the underlying soil (pervious/impervious) and C_r has a recommended value of 1.3, so the overall effect is probably not large. Also changed somewhat is the time of entry, which is related to length and slope of the sub-catchments and varies from 3 to 10 minutes depending on the chosen design frequency, instead of the 2 minutes of the TRRL.

10.3 Hydrograph methods

The Rational Method described provides an estimate of peak flow in a system but does not provide a discharge hydrograph. Nor does it accommodate the variation in rainfall intensity with time, which occurs naturally.

The first successful attempt at modelling a sewerage system by hydrographs based on rain profiles, rather than a uniform average intensity as assumed in the Rational Method, was the TRRL Hydrograph method which is still widely used. However, the Hydrograph method of the Wallingford Procedure is considerably more complex and, designed from the outset for use as a series of computer programs, it represents the existing state of the art. Its basis is as follows:

(a) Design rainfall profiles are calculated for specified values of duration, return period, M5–60 minutes and r values for a particular location (see chapter 2, section 2.9.1). Summer profiles (as distinct from the Winter profiles of section 2.9.5) should be used for periods of 15, 30, 60 and 120 minutes, for each return period examined.

(b) The data of the pipe network and the contributing areas, together with C_v and C_r values, are entered.

(c) The percentage runoff from the whole catchment is calculated and separate runoff factors from the three surface types (paved surfaces, pitched roofs and pervious areas) are calculated.

(d) The chosen rainfall hyetograph from (a) is converted into ten standard runoff hydrographs using the surface runoff models for three characteristic slopes and three types of surface plus one for pitched roofs. Depression storage (if it exists) is deducted from the hyetograph first.

(e) The surface runoff hydrograph which contributes directly to the pipe is calculated from the standard hydrographs, the contributing areas and the percentage runoff factors.

(f) The hydrograph so obtained is added to the inflow hydrograph at the upstream manhole.

(g) The program then determines the smallest standard pipe size, of specified shape, to convey the peak discharge at the upstream manhole.

(h) The combined hydrograph is routed through the pipe using the Muskingum–Cunge routine model for free-surface flow.

(i) Steps (e)–(h) are repeated for each pipe length.

(j) Steps (e)–(i) are repeated for a series of rainfall events of the same return period but differing duration.

The Wallingford procedure includes further programs for optimising a proposed system and simulating an existing system. Ancillary sewerage structures, storage ponds and combined storm water and foul sewerage systems can all be dealt with.

References

1. LLOYD-DAVIES, D. E. The elimination of storm-water from sewerage systems. *Proc. Inst. Civ. Eng.*, **164** (1906) 41
2. *TRRL Road Note 35.* A guide for engineers to the design of storm sewer systems, 2nd edn, HMSO, London, 1976
3. Hydraulics Research Station. *Charts for the hydraulic design of channels and pipes*, 4th edn, HMSO, London, 1978
4. Dept of Environment. *The Wallingford Procedure Design and analysis of urban storm drainage* (5 volumes), HMSO, London, 1981

Further reading

COLYER, P. J. Performance of storm drainage simulation models. Proc. Inst. Civ. Engrs., **63**, Part 2 (June 1977) 293

Construction Industry Research and Information Association, Rainfall, runoff and surface water drainage of urban catchments. *Proceedings of the research colloquium at Bristol, April 1973*, published Nov. 1974

FORD, W. The adaption of the RRL Hydrograph Method for tropical conditions. Proceedings of Symposium on Flood Hydrology, Nairobi, 1975. *Transport and Road Research Laboratory, Supplementary Report 259*, 1977

KIDD, C. H. R. and PACKMAN, J. C. Selection of design storm and antecedent condition for urban drainage design. *Report No 61*, Institute of Hydrology, 1979

MAYS, L. W. and YEN, B. C. Optimal cost design of branched sewer systems. *Water Resources Research*, 11, No. 1 (1975) 37

PACKMAN, J. C. and KIDD, C. H. R. A logical approach to the design storm concept. *Water Resources Research*, 16, No. 6 (1980) 994

van den BERG, J. A. Data analysis and system modelling in urban catchment areas (in the New Town of Lelystad, The Netherlands). *Hydrol. Sciences Bulletin XXI*, 1, No. 3 (1976)

YEN, B. C. *et al.* Advanced methodologies for design of storm sewer systems. *Water Resources Centre Report No 112*, University of Illinois, 1976

11 International Flood Frequency Growth Curves

11.1 World Flood Studies

Since the Flood Studies Report was published, many of the techniques devised for it have been subsequently developed and their range of application extended. One of the investigations arising from this development was the World Flood Study [1] which, in the words of its authors, "was conceived with the aim of examining and classifying the characteristics of floods in as many countries and from as wide a range of climates as possible." Much of this work has been subsequently reported in the literature [2].

Data were collected from individuals and organisations around the world in 70 countries, and were assessed and analysed uniformly to allow comparisons of results between stations and countries. Both annual maximum flood series and catchment characteristics were collected.

The results were used to produce growth curves (see figure 9.8) showing the increasing value of the parameter $Q/\bar{Q}$ with decreasing frequency of occurrence. By making the curves non-dimensional in this way, results could be compared by region, climate, catchment area and annual rainfall, for homogeneity.

It was, of course, necessary to establish $\bar{Q}$, the mean annual flood (termed MAF in [1, 2]) from the data set available in each region or country, but the growth curves derived can be used to estimate floods at sites where there are no gauging stations, always provided that $\bar{Q}$ may be estimated from catchment characteristics.

A selection of the growth curves is shown in figures 11.1 to 11.4. It is emphasised that these represent averages of broad regions and, for specific projects and detailed design, more data would always be necessary. The curves presented in the figures are only some of those available. For example, Thailand on figure 11.4 has five curves representing different altitudes and catchment areas, the one shown as no. 5 being an average curve for the country, excluding the Malay peninsula. Those countries marked with an asterisk also have other more rigidly-defined catchment curves presented in [1]. North and South America feature in the source publication but are not sampled here.

Interested readers are recommended to consult the original source publications for details of the data used, the statistical techniques employed, and the limitations and qualifications about the use of the curves.

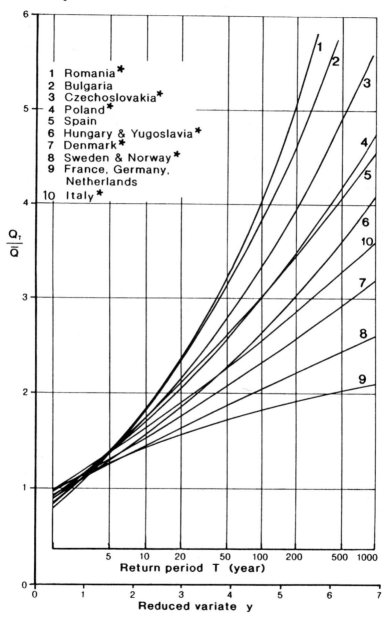

Figure 11.1 *European flood frequency curves*

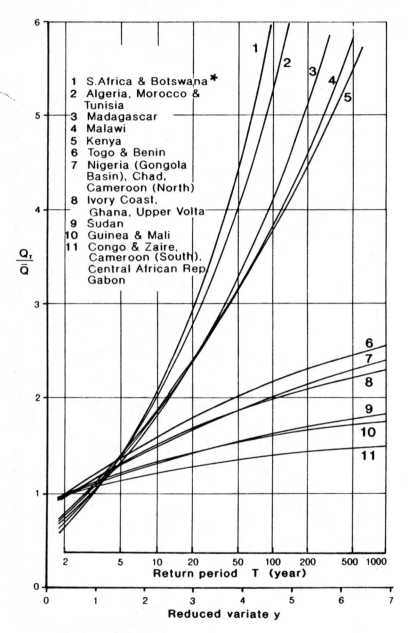

Figure 11.2 *African flood frequency curves*

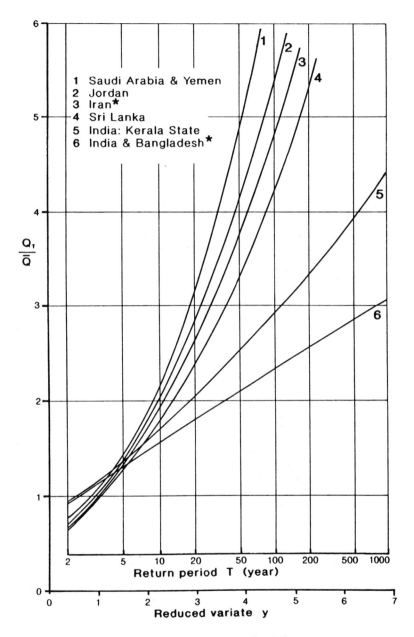

Figure 11.3 *Middle Eastern and Indian flood frequency curves*

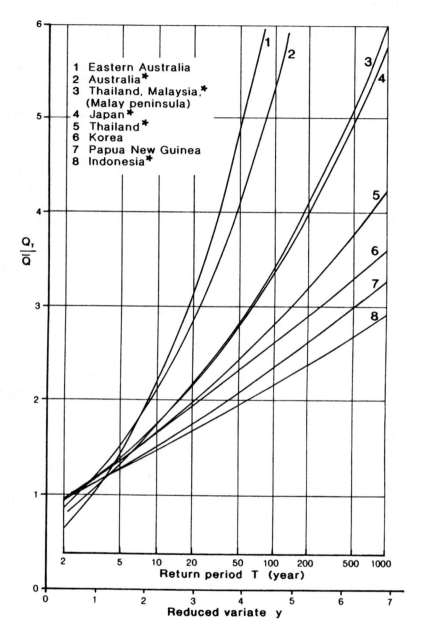

Figure 11.4 *Asian and Australasian flood frequency curves*

11.2 General conclusions

Certain trends are evident from the material shown here. For example, semi-arid climates have the steepest growth curves since the infrequent heavy rainfall produces floods relatively much greater than the mean annual flood. Examples of this in figure 11.2 are the curves for South Africa, Algeria, Morocco and Tunisia, and in figure 11.3, Saudia Arabia and Jordan. Similarly, the equatorial climates with high annual rainfall tend to have flatter curves, e.g. the Nigeria, Ghana, Congo group in figure 11.2 and Indonesia and Papua New Guinea in figure 11.4.

Clearly, the similarity of climate and topography are not the only factors influencing the growth curves since Sri Lanka and Kerala State in India are both subject to the South West monsoon but have appreciably different curves.

Other important factors are area, soils and type of vegetation. The larger the catchment area, the flatter the curve is a general rule, though the authors of [1] mention one exception to this in the Malay peninsula.

The Institute of Hydrology in the UK is continuing to expand the database of flood series and catchment characteristics, and invites interested engineers and hydrologists throughout the world to send it such information.

References

1. MEIGH, J. and FARQUHARSON, F. A. K. *World Flood Study, Phase II.* Institute of Hydrology, Wallingford, UK, Nov. 1985
2. FARQUHARSON, F. A. K. *et al.* Comparison of flood frequency curves for many different regions of the World, from *Regional Flood Frequency Analysis* (V. P. Singh, ed.), Reidel, Dordrecht, 1987

Further reading

RODIER, J. A. and ROCHE, M. World catalogue of maximum observed floods. *Int. Assoc. Hydrological Science*, No. 143 (1984)
UNESCO. *World catalogue of very large floods*, UNESCO Press, Paris, 1976
WILTSHIRE, S. E. Grouping basins for regional flood frequency analysis. *Hydr. Sci. Journal*, **30** (1985) 151

12 Design Criteria

12.1 Risk analysis

All hydraulic engineering design carries with it an implied structural life during which the structure is expected to meet its design specification. Some structures have specified design lives (for example, a cofferdam or a diversion tunnel) but often such a life is not made explicit. How long, for example, are the design lives of a canal, a dam, or a breakwater?

The engineering designer has to approach the problem of design of hydraulic structures from consideration of the damage that would ensue if failure of the structure occurred and/or if it is possible to exceed the design specification without structural failure. It is not sufficient to consider return period alone; what is needed is an understanding of the risk of encountering certain conditions during specific periods of time, and of the consequences of the design flow being exceeded, including the danger to life and the economic, environmental and social effects of failure of the structure. Only then can a proper judgement be formed as to whether or not enough has been done to make the risk acceptable without economic repercussions that are unacceptable.

It may be difficult to make any quantitative assessment of these factors, but consideration of a series of steps may at least provide a logical means of comparison of the various options open to the designer:

(a) Identification of the events or sequences of events which may lead to failure, and determination of their probability of occurrence.
(b) Identification of specific features of the structure which might initiate failure or partial failure (e.g. failure of gates to operate or be operated, loss of power, etc.) and estimates of their probability.
(c) The likelihood of combinations of events in (a) and (b).
(d) The consequences, including the forecast economic, social and environmental costs of each combination in (c).

The thoroughness with which these studies are made will depend on the nature of the structure. Dams are the most immediately obvious risks, but canals and their control structures, river bank revetments and sea defence works are all hydraulic structures liable to need such risk analysis.

In the end, there must be engineering judgement about the acceptable level of risk. Such judgement can best be exercised using whatever quantification of the risk can be made, and it should always be made clear to the owner/client concerned what is the basis for a recommended level of installation. A much fuller discussion of the subject has been published in references [1] and [2].

The discussion of damage, and how it can be described and assessed in this context, is beyond the scope of this text, but has been studied by Borgman [3], US Army Corps of Engineers [4] and Penning-Rowsell and Chatterton [5].

12.2 Choice of design return period by consideration of design life and probability of encountering design flow during this life

In table 9.1, the probability of the N-year flood occurring in a specific period of years is set out. Table 12.1 is a somewhat expanded version of table 9.1, and a different terminology is employed to make the concept clear. The return period of an event (T_r) is related to the arbitrarily chosen design life (L) to give the probability of encountering the event, or something greater, in that design life (P). The table is based on the expression

$$P = 1 - \left(1 - \frac{1}{T_r}\right) L$$

TABLE 12.1 *Percentage risk of encountering an event within a particular design life L for different return periods[a]*

Design life L (years)	Return period T_r (years)								
	5	10	20	30	50	100	200	500	1000
1	20	10	5	3	2	1	—	—	—
2	36	19	10	7	4	2	1	—	—
3	49	27	14	10	6	3	1	—	—
5	67	41	23	16	10	5	2	1	—
7	79	52	30	21	13	7	3	1	1
10	89	65	40	29	18	10	5	2	1
15	96	79	54	40	26	14	7	3	1
20	99	88	64	49	33	18	10	4	2
30	—	96	78	64	45	26	14	6	3
50	—	99	92	82	64	39	22	9	5
75	—	—	98	92	78	53	31	14	7
100	—	—	99	97	87	63	39	18	10
150	—	—	—	99	95	78	53	26	14
200	—	—	—	—	98	87	63	33	18
300	—	—	—	—	—	95	78	45	26
500	—	—	—	—	—	99	92	63	39
1000	—	—	—	—	—	—	99	86	63

[a]Where no figure appears the risk is either less than 0.5 per cent or more than 99.5 per cent

12.3 Choice of a design value of a rare event

Many of the techniques described in this text were developed to provide the designer with numerical values of discharge, rainfall etc. that have physical meaning (that is, they yield velocity, depth, volume and so on).

The various numbers that result from calculations, however, will not all have equal significance because they have been derived in different ways, and care needs to be taken before they are used as specifications. Consider the following three examples.

(a) *Probable maximum precipitation.* Although in section 2.10 ways are discussed of arriving at this figure, it is questionable whether, in the British Isles, there is any need to look further than the boundary equation of figure 2.4, since the data are so extensive and the record so long. In areas less well-documented and observed the more fundamental approach may be appropriate.

(b) *The Q_T value derived from the FSR equations (and their derivatives).* The equations for $\bar{Q}$ were derived by multiple regression of catchment parameters and are equations of best-fit. This means that there will have been as many underpredictions as overpredictions. The designer who wishes to be confident that *his* catchment will not be one that is underpredicted should apply a factor of safety (F_n) [6]. A value of F_n = 2.26 will give 95 per cent confidence and F_n = 1.26 will give 68 per cent confidence that $\bar{Q}$, and hence Q_T, is not underpredicted; 98 per cent confidence would require F_n = 2.7 [7]. Clearly it is a matter of engineering judgement what value of F_n should be used, since there is already an element of safety-factoring in the choice of T for Q_T.

(c) *The Q_T value derived from a long period of record.* The recorded annual maxima of the River Thames at Teddington (table 9.2) are likely to provide a value, however analysed, of (say) Q_{200} with an integrity that an ungauged catchment cannot provide. In such an instance a factor of safety applied to the Q value is probably inappropriate, and a conservative value derived from a variety of plotting methods may well suffice.

References

1. Evaluation procedures for hydrologic safety of dams. *Report by Task Committee on Spillway Design Flood Selection*, Hydraulics Division, ASCE, 1988
2. Guidelines to Decision Analysis. *ACER Technical Memorandum No. 7*, Denver, USA
3. BORGMAN, L. E. Risk criteria. *J. Waterways and Harbours Division, ASCE,* **89** (1963)
4. *Expected Annual Flood Damage Computation*, US Army Corps of Engineers, Hydrologic Eng. Centre, 1977

5. PENNING-ROWSELL, E. C. and CHATTERTON, J. B. *The benefits of flood alleviation: A manual of assessment techniques*, Saxon House, 1977
6. POOTS, A. D. and COCHRANE, S. R. Design flood estimation for bridges, culverts and channel improvement works on small rural catchments. *Proc. Inst. Civ. Eng.*, **66**, TN 229 (1979) 663
7. Ref. 6, discussion by M. A. BERAN, *Proc. Inst. Civ. Eng.*, **68** (1980) 319

Further reading

MOSONYI, E. and BUCK, W. Some aspects of risk analysis for improved water resources planning. *Proc. 2nd World Congress Int. Wat. Res. Assoc. New Delhi 1975*, Vol. 2, p. 221

Appendix A Rainfall and Soil Characteristics of the British Isles

The British and Irish National Grids are illustrated. The rainfall and soil type maps are at a scale such that there are ten separate sections covering Britain (with the exception of some Scottish islands) and four covering Ireland.

Each section has national grid lines identified by numbers signifying hundreds of kilometres from the origin and is itself identified by a large bold number (for example, southeast England is **3**). For convenience, four different maps of each section are kept together. These are:

(a) standard average annual rainfall (hundreds of mm)   SAAR
(b) 2-day M5 rainfall (mm)   2DM5
(c) ratio *r* = 60-minute M5/2-day M5 (as a percentage)   *r*
(d) soil classification for runoff potential   RP
    (see section 9.4 for calculation of SOIL term)

Each map has one of these abbreviated titles followed by a bold number that identifies the section on the key map. Britain is dealt with first and then the Irish key map and Irish sections follow.

These maps are reproduced from the *NERC Flood Studies Report*, 1975, except for RP which is from the 1978 revision of *Winter Rain Acceptance Potential*, published as Supplementary Report No. 7.

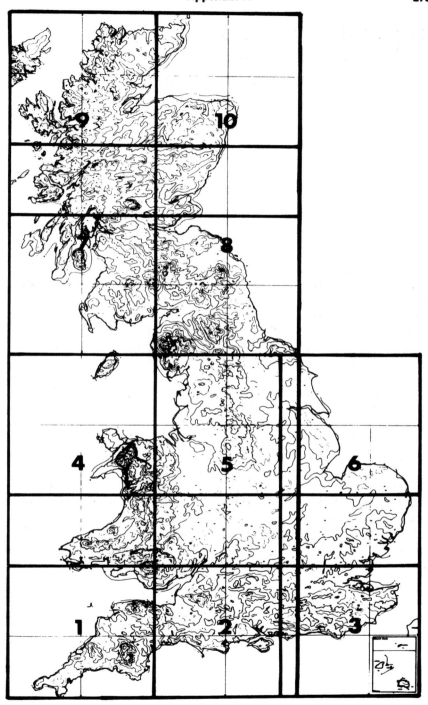

Key to section numbers for maps of Great Britain

Diagram showing 100 km squares and the letters
used to designate them

British National Grid

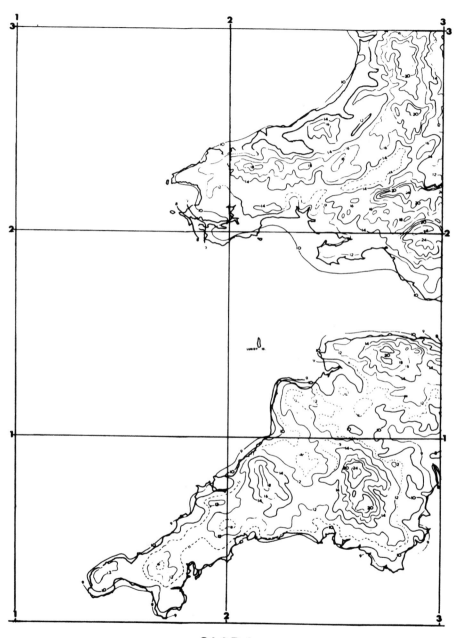

SAAR.1

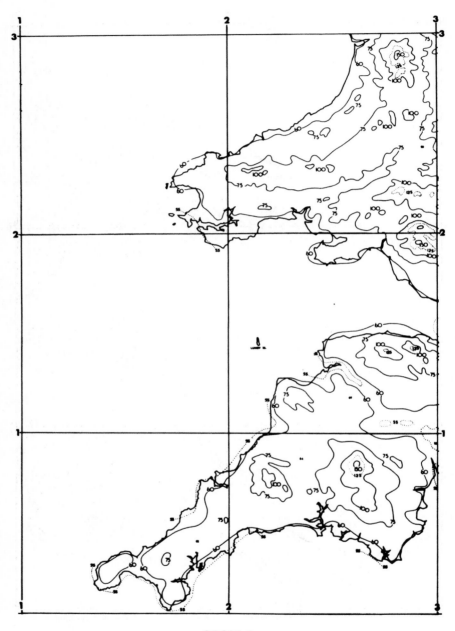

2DM5.1

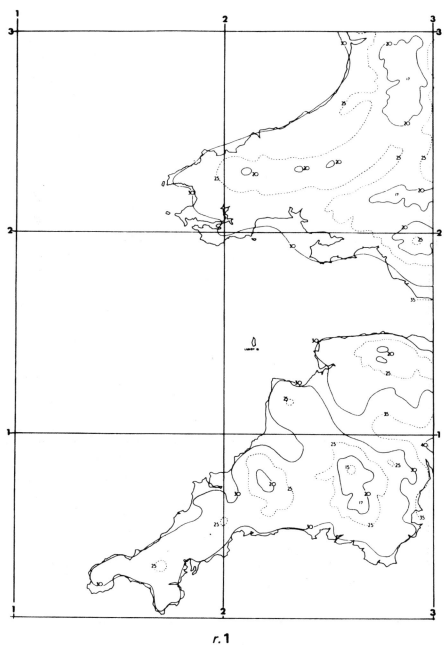

r. **1**

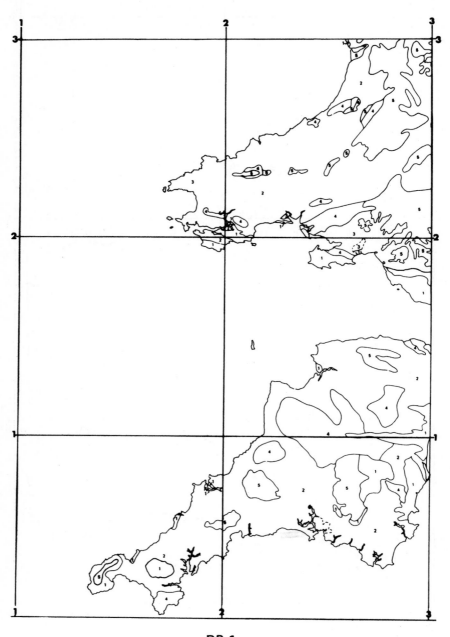

RP.1

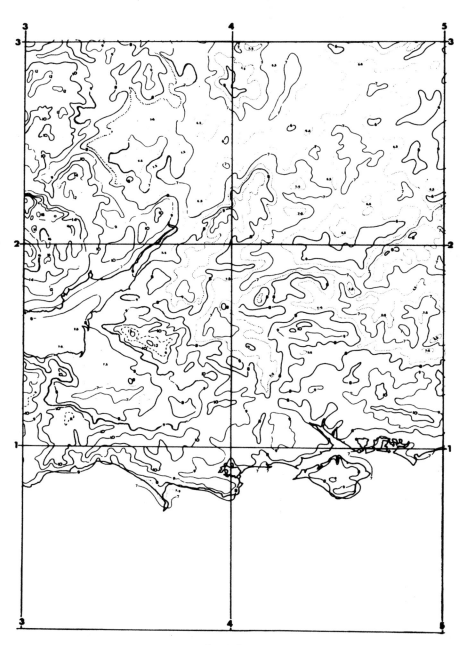

SAAR.2

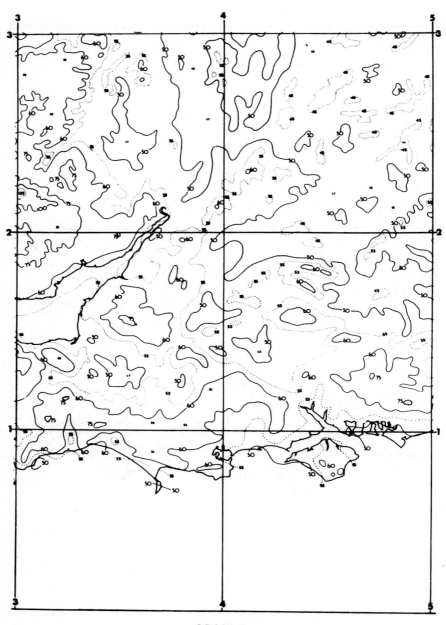

2DM5.2

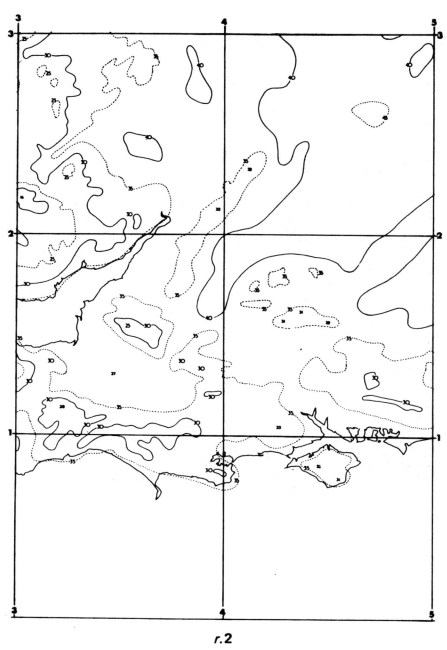

*r.*2

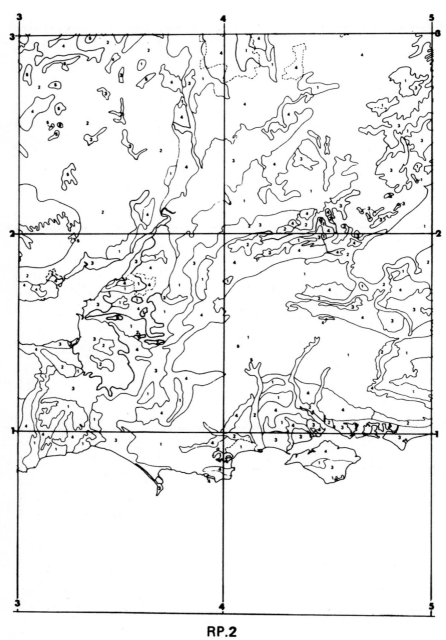

RP.2

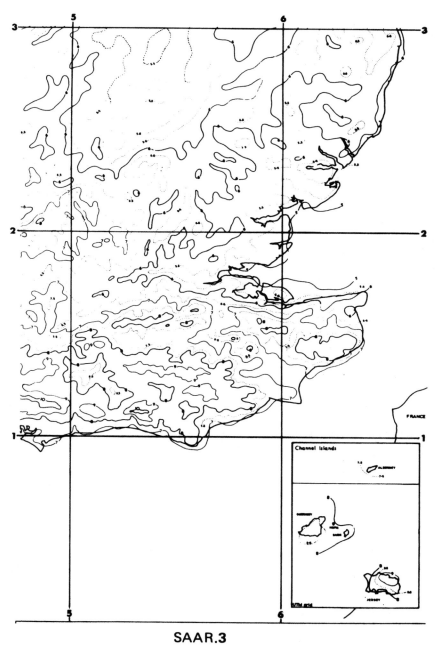

SAAR.3

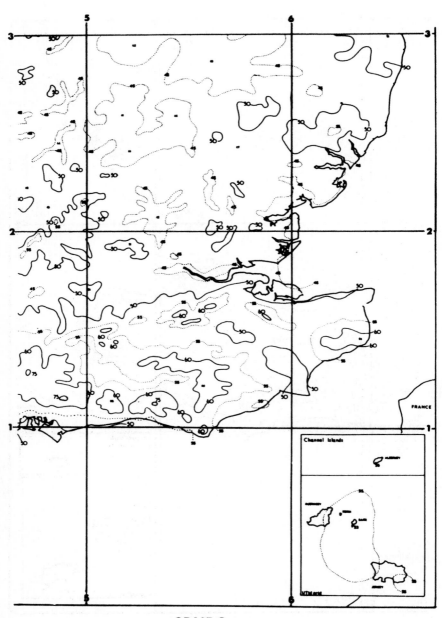

2DM5.3

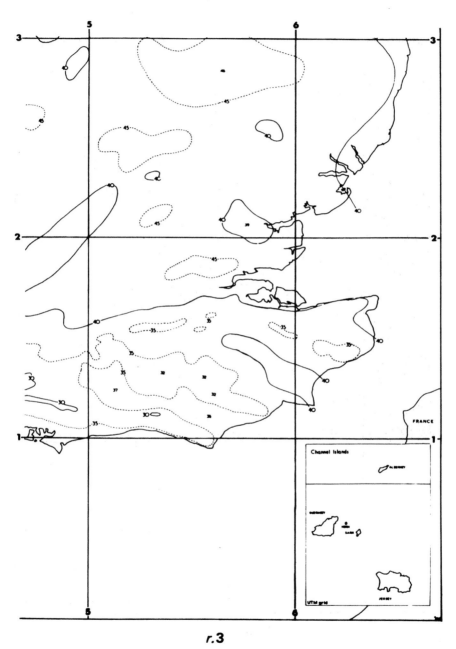

r.3

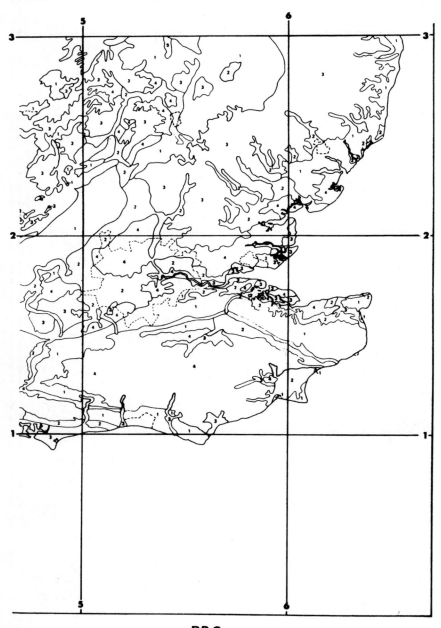

RP.3

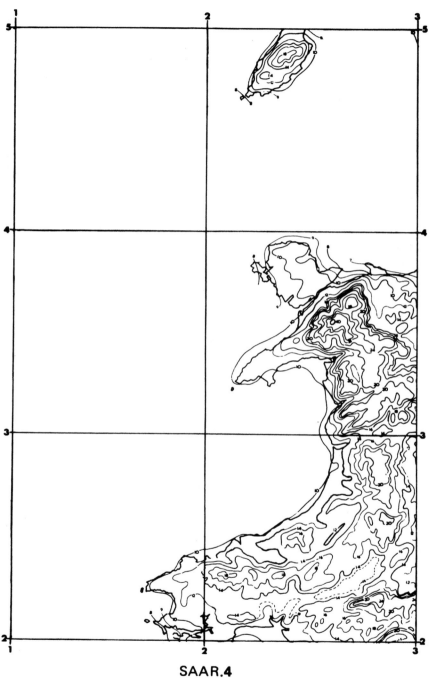

SAAR.4

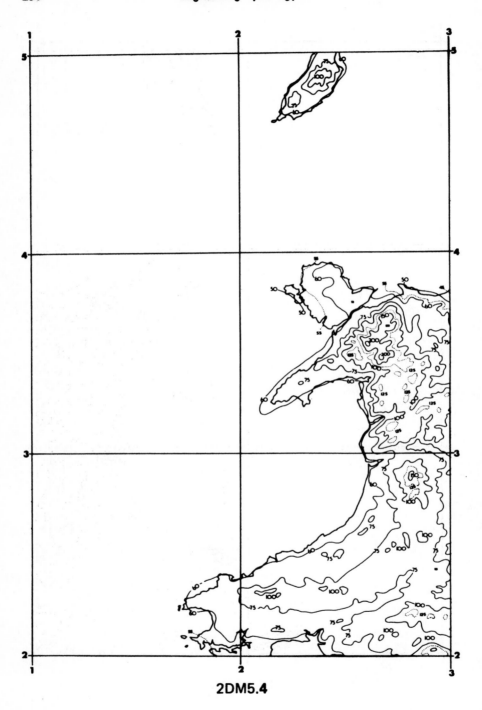

2DM5.4

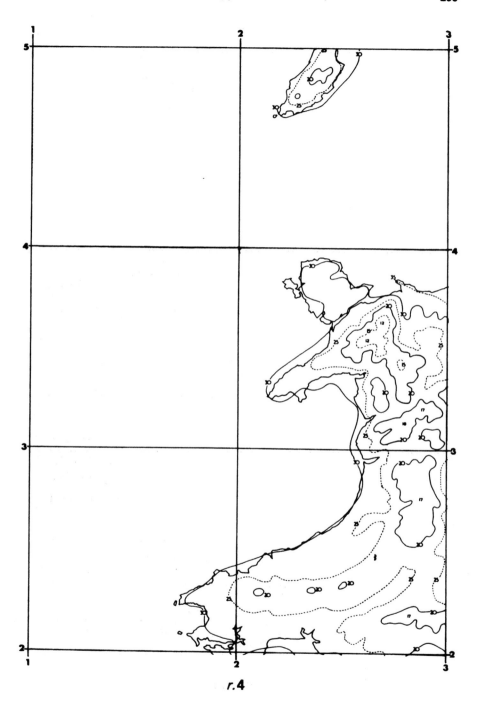

r.4

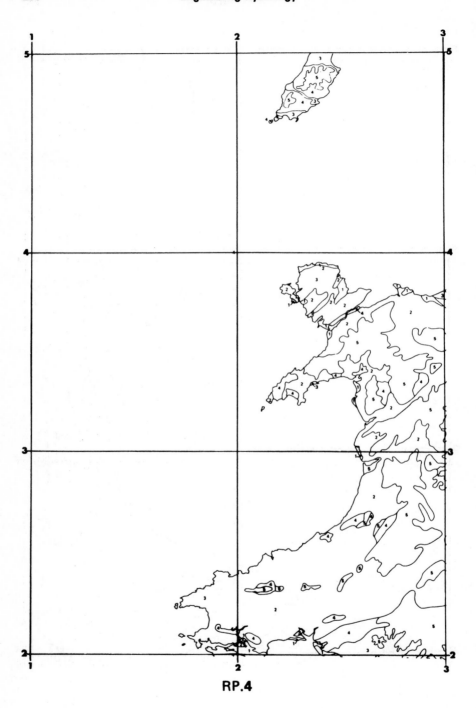

RP.4

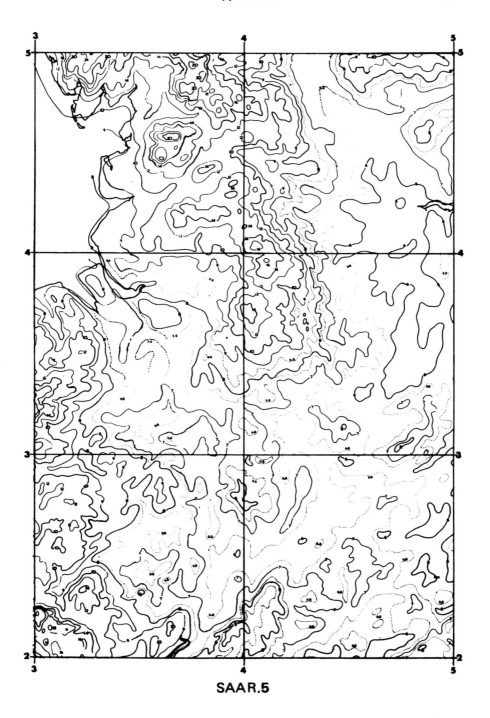

SAAR.5

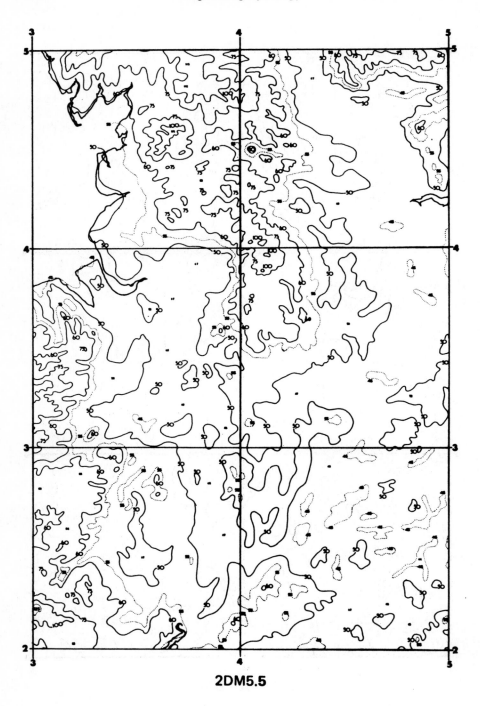

2DM5.5

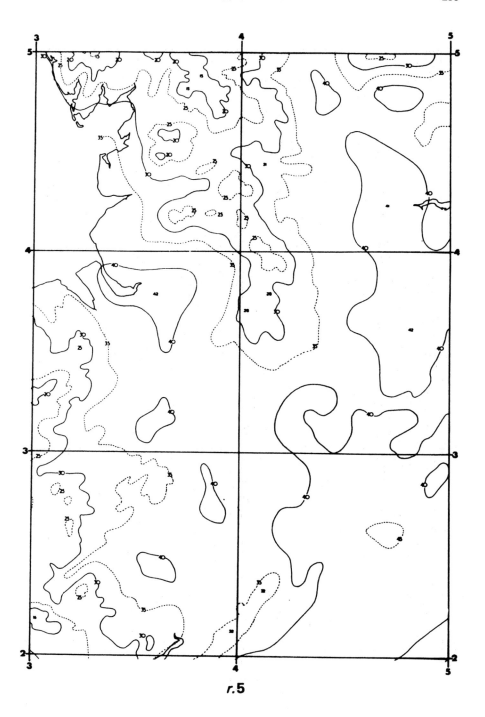

*r.*5

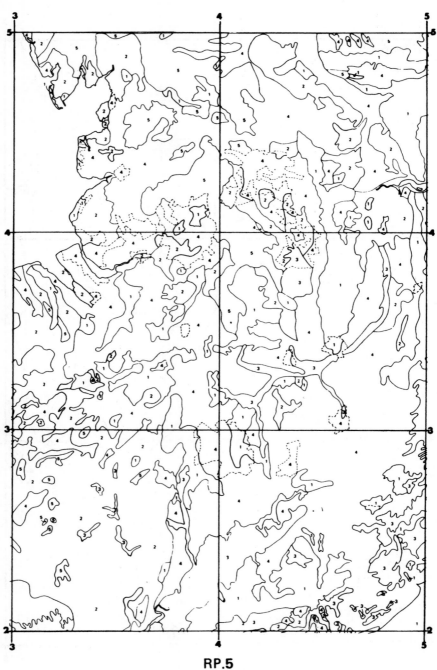

RP.5

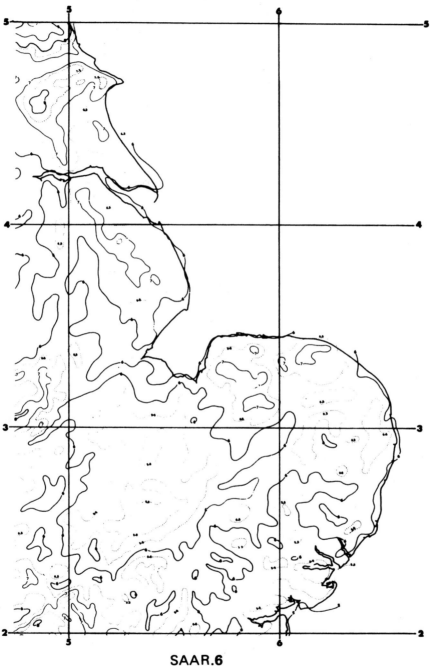

SAAR.6

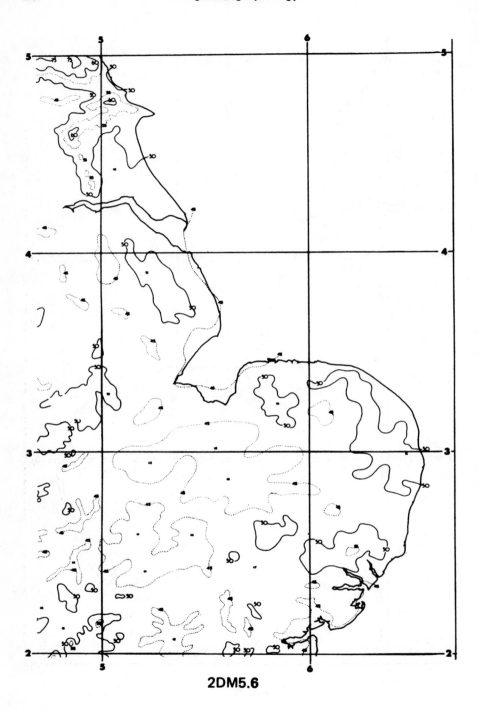

2DM5.6

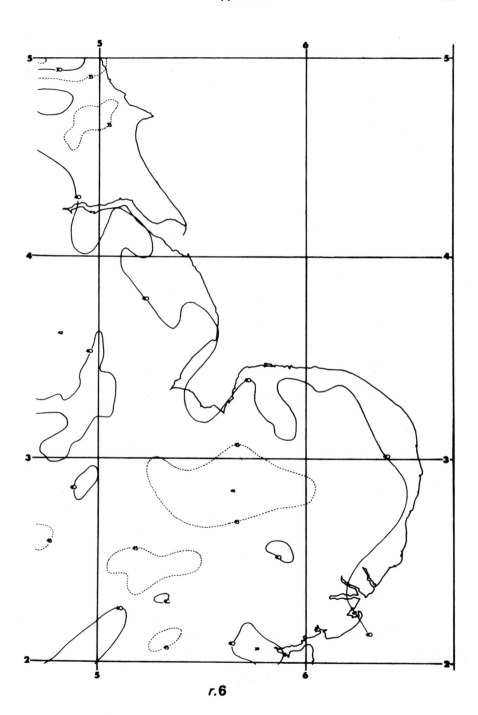

*r.*6

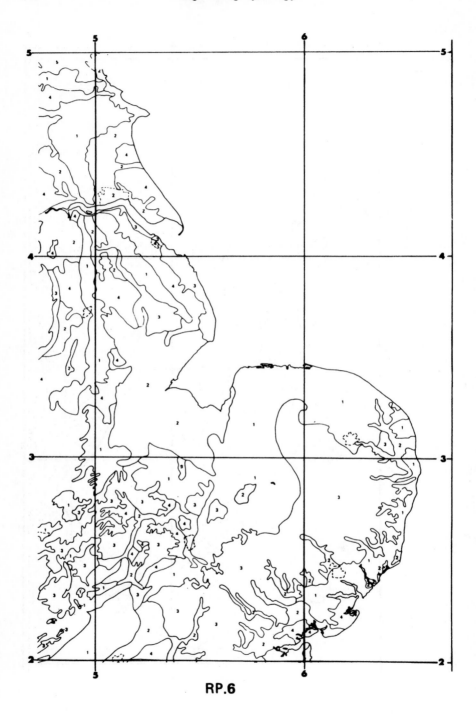

RP.6

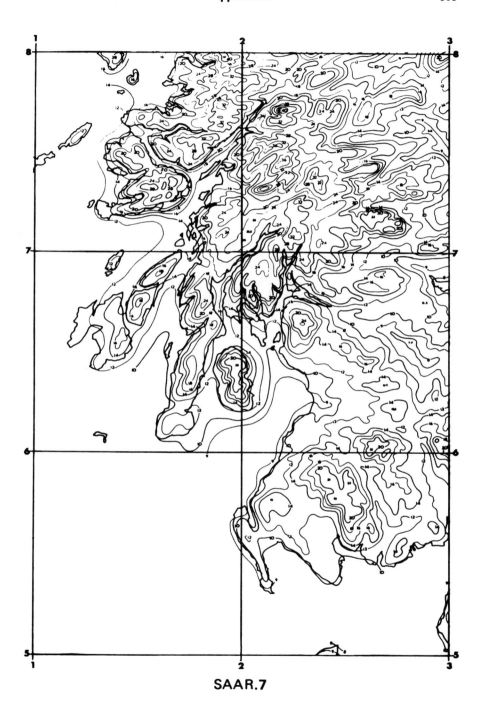

SAAR.7

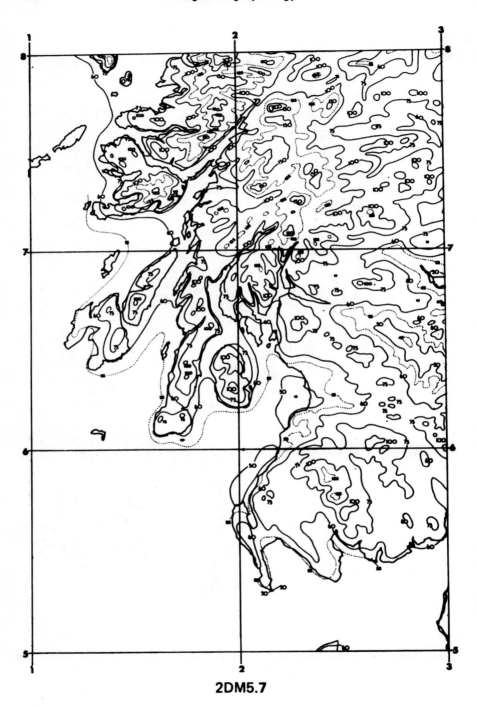

2DM5.7

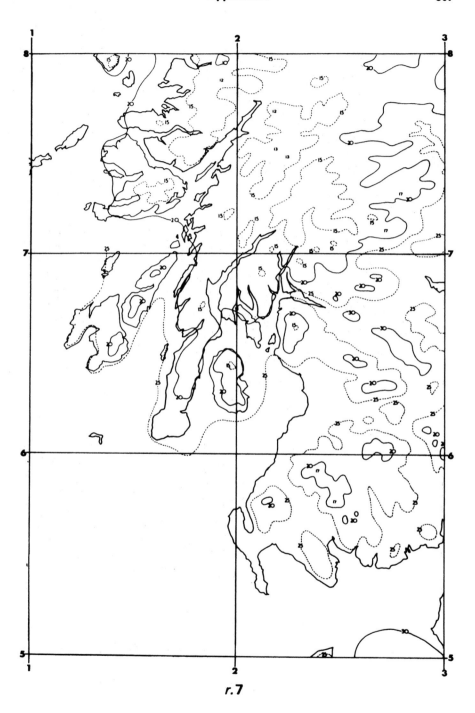

r.7

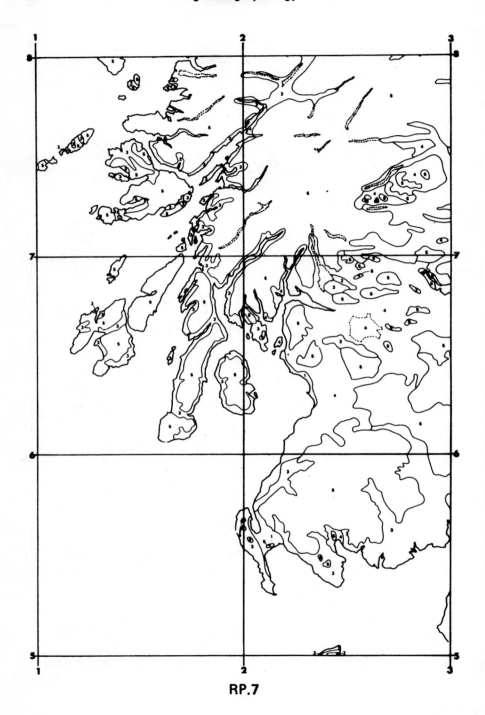

RP.7

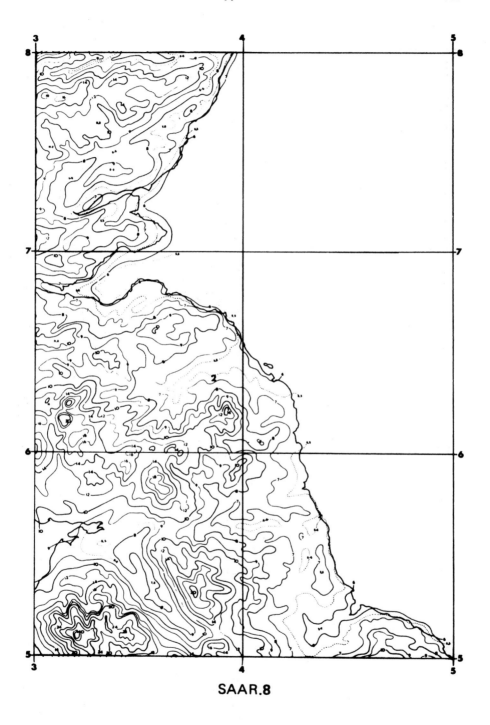

SAAR.8

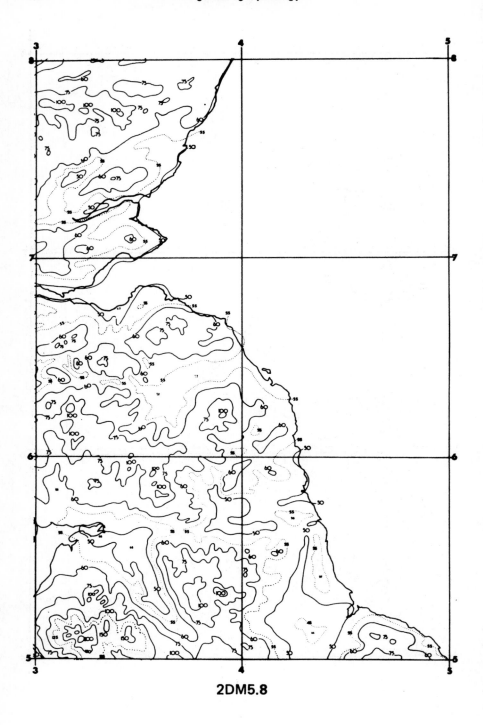

2DM5.8

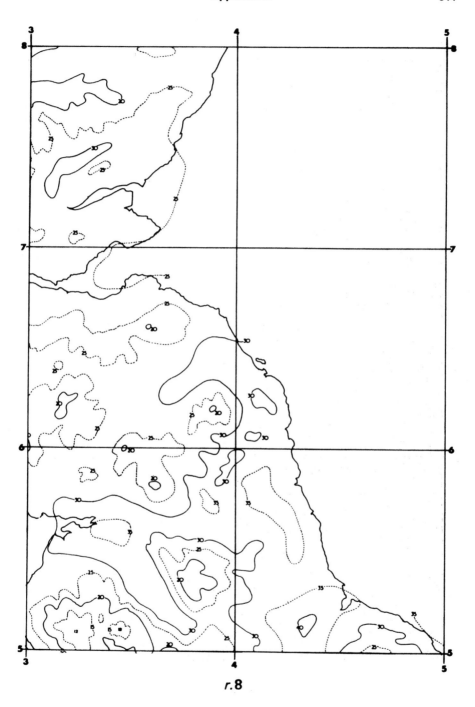

r.8

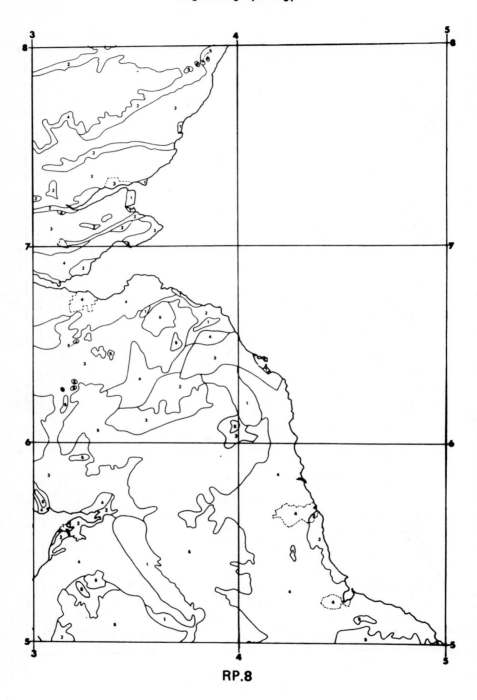

RP.8

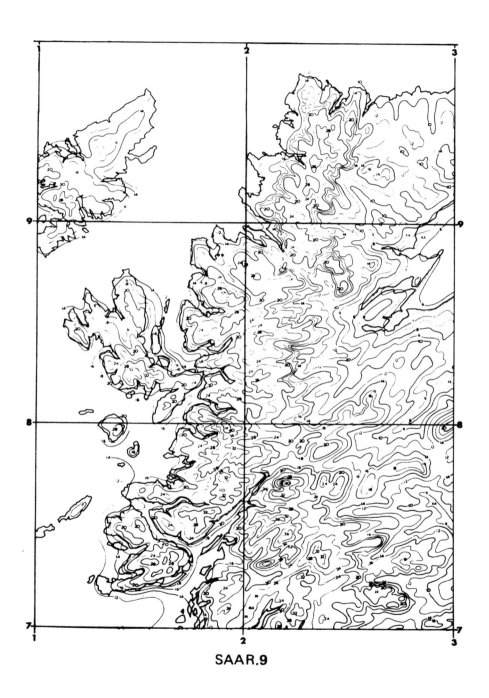

SAAR.9

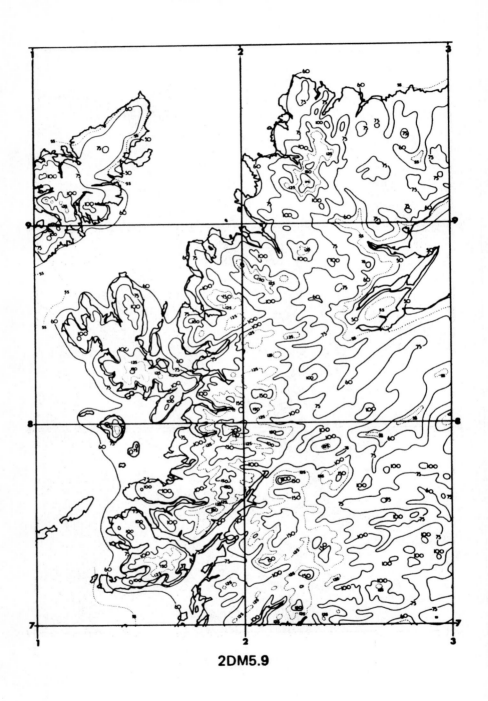

2DM5.9

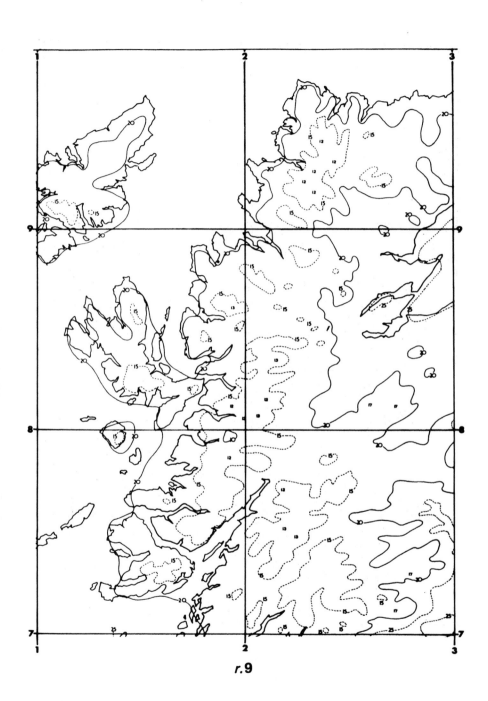

*r.*9

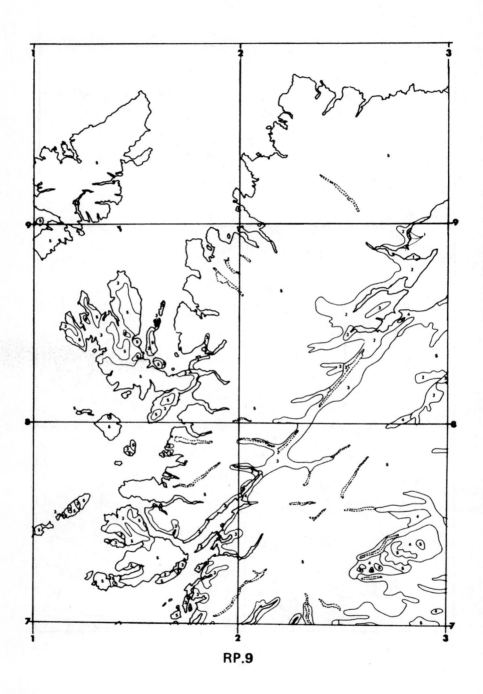

RP.9

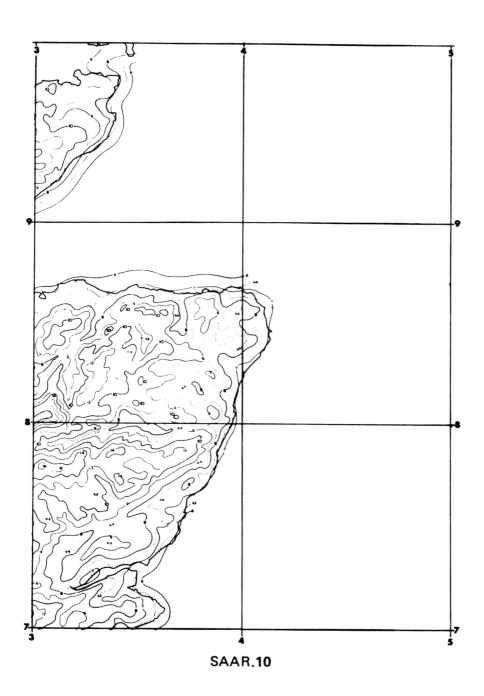

SAAR.10

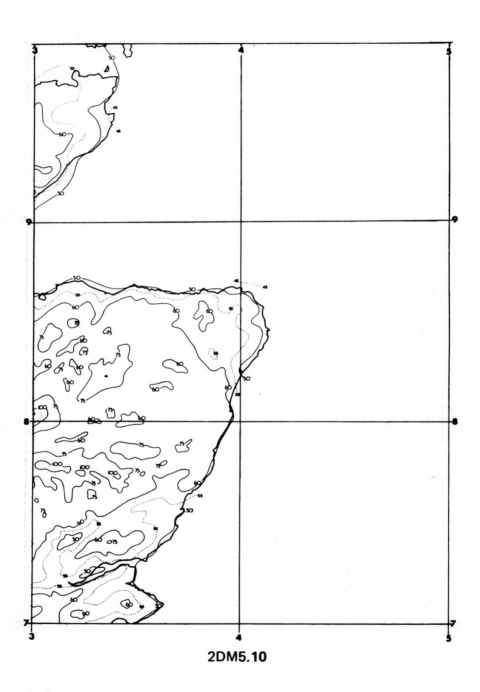

2DM5.10

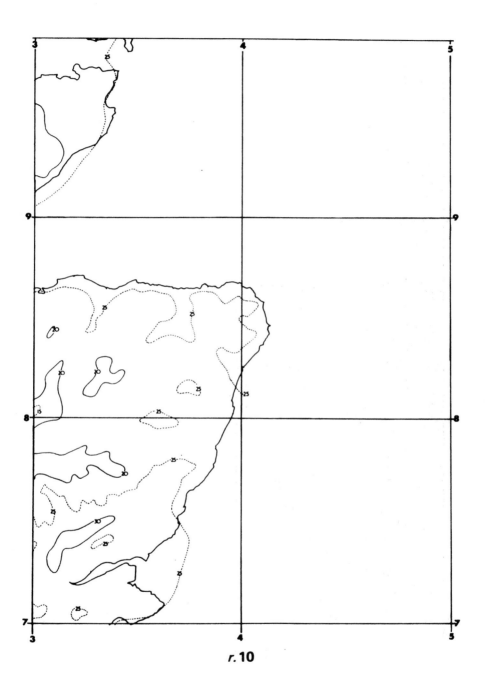

r. 10

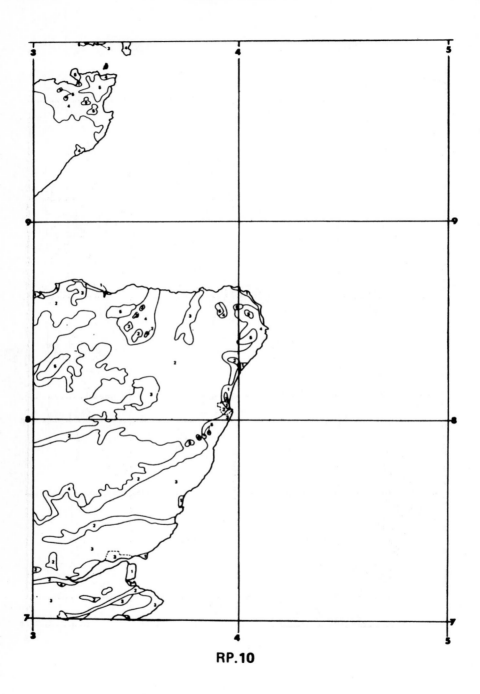

RP.10

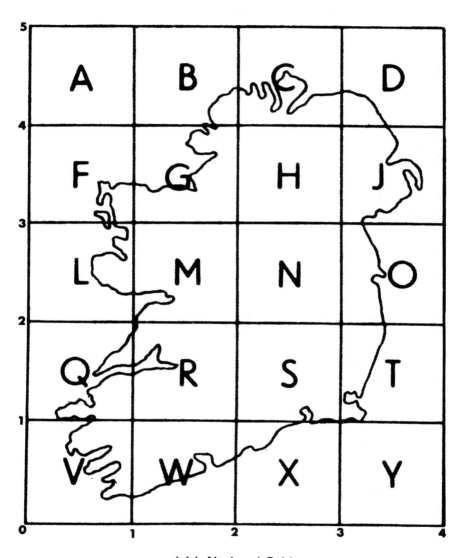

Irish National Grid

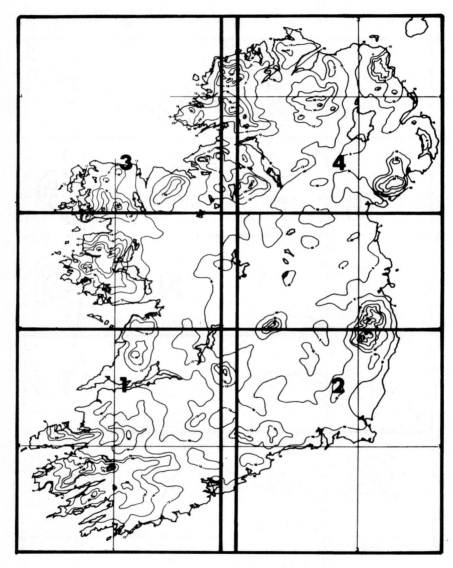

Key to section numbers for maps of Ireland

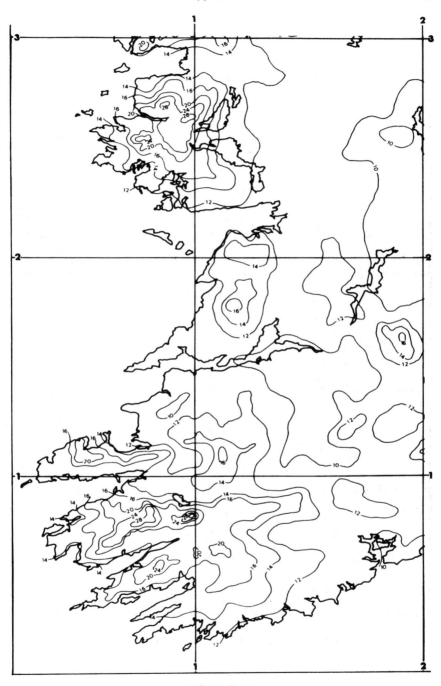

I SAAR.1

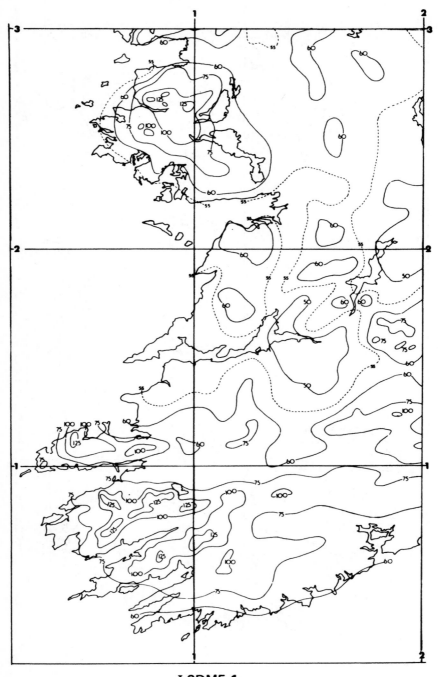

I 2DM5.1

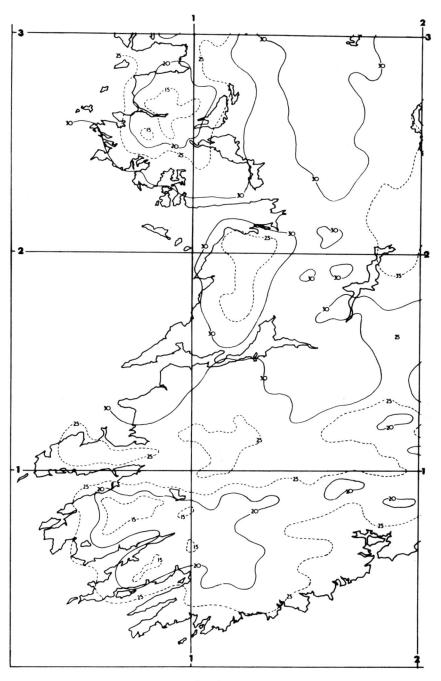

I *r*.1

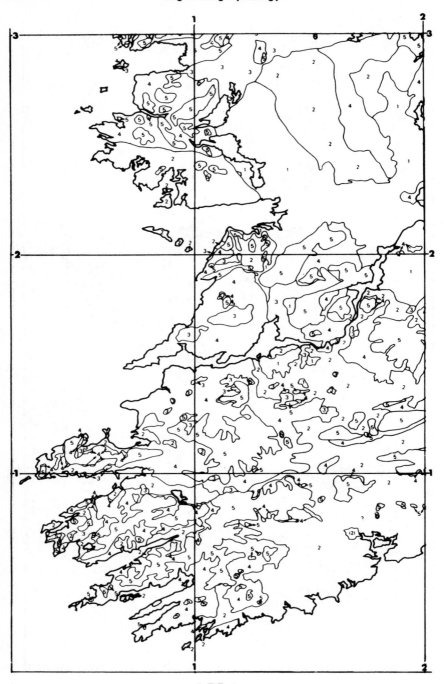

I RP.**1**

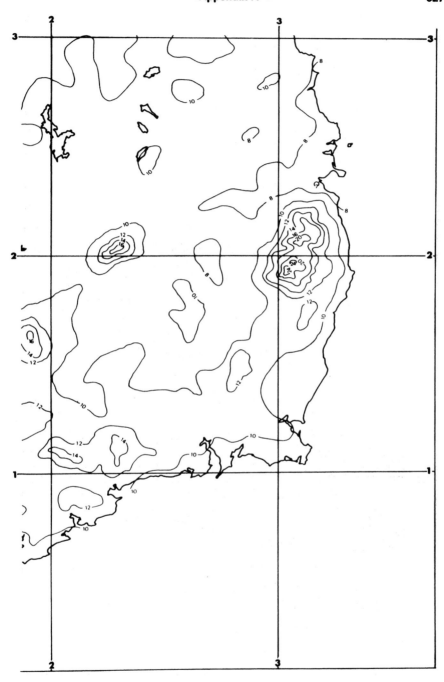

I SAAR.2

I 2DM5.2

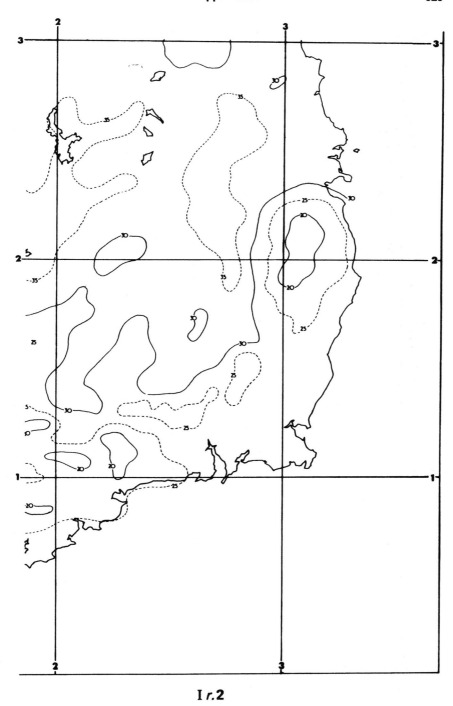

I r. 2

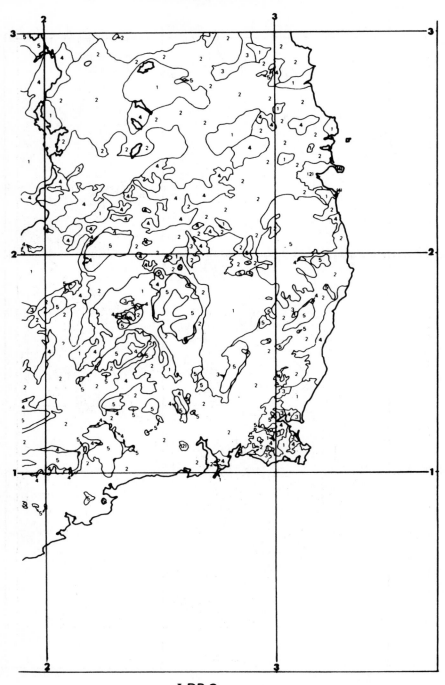

I RP.2

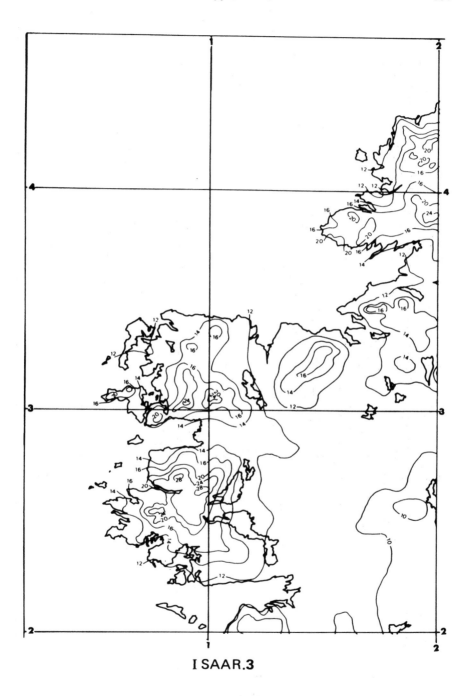

I SAAR.3

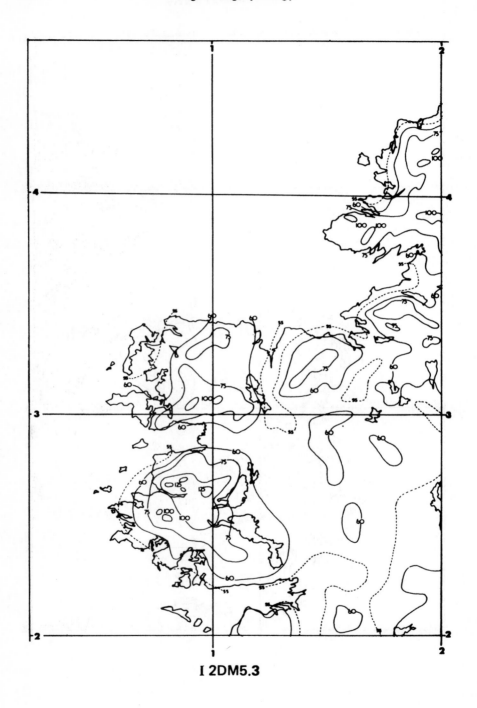

I 2DM5.3

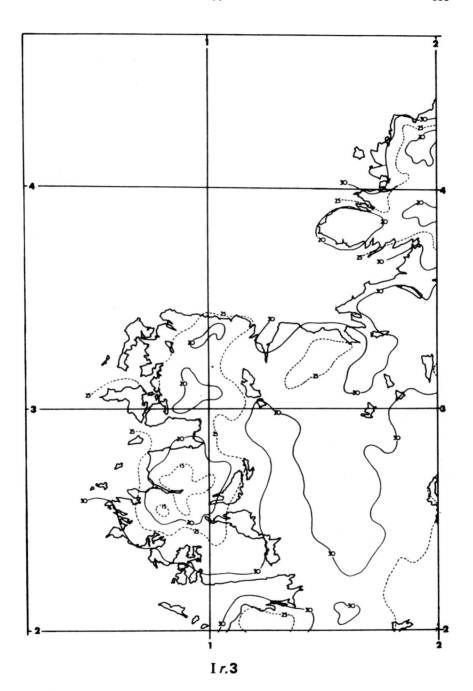

I r.3

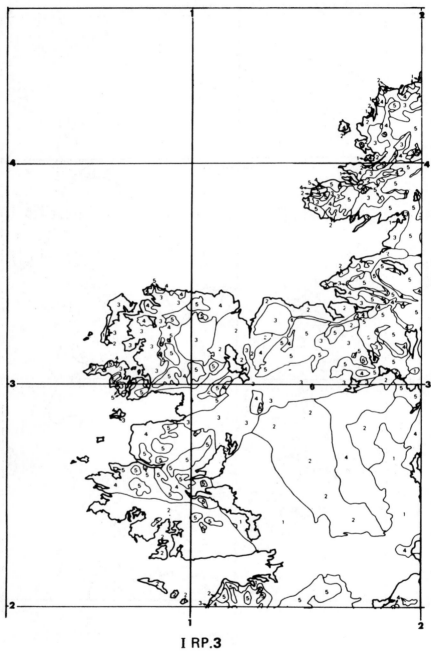

I RP.3

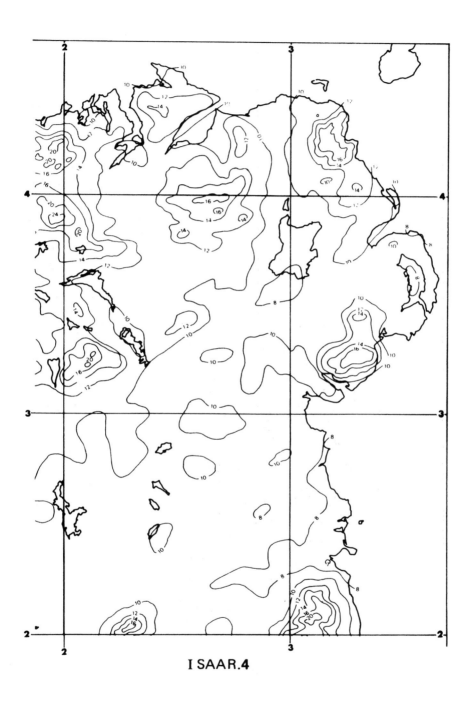

I SAAR.4

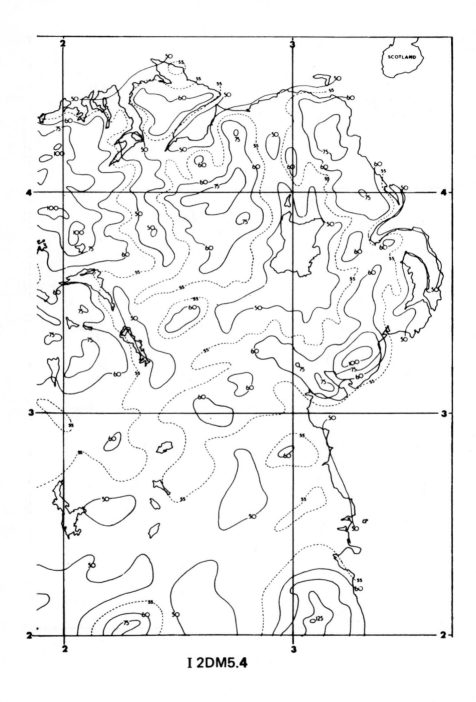

I 2DM5.4

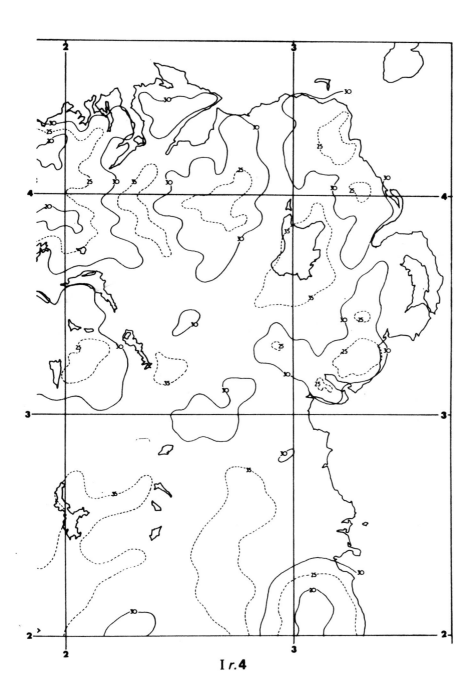

I r.4

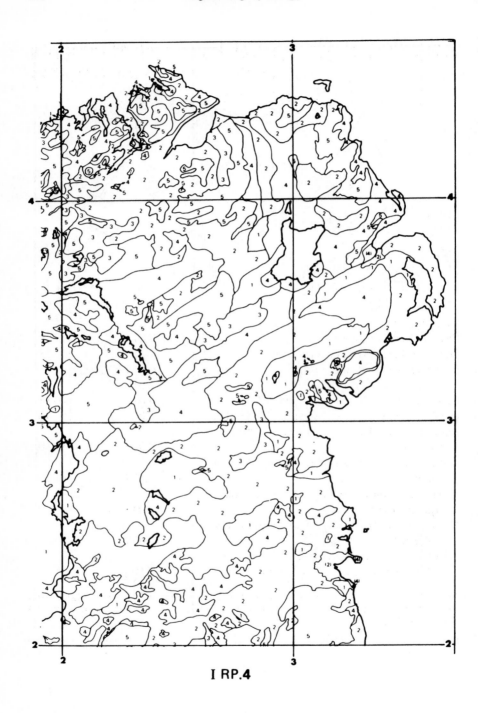

I RP.4

Appendix B Typical Values of Manning's n in $Q = (1/n)(AR^{2/3} S^{1/2})$ and Chezy's C^* in $V = C\sqrt{(RS)}$

Type of channel	n	C (SI units)
Smooth timber	0.011	
Cement-asbestos pipes, welded steel	0.012	70-90
Concrete-lined (high-quality formwork)	0.013	60-75
Brickwork well-laid and flush-jointed	0.014	
Concrete and cast iron pipes	0.015	
Rolled earth: brickwork in poor condition	0.018	40-55
Rough-dressed-stone paved, without sharp bends	0.021	30-45
Natural stream channel, flowing smoothly in clean conditions	0.030	19-30
Standard natural stream or river in stable condition	0.035	14-25
River with shallows and meanders and noticeable aquatic growth	0.045	
River or stream with rocks and stones, shallow and weedy	0.060	
Slow flowing meandering river with pools, slight rapids, very weedy and overgrown	0.100	

* For a full discussion of Chezy's coefficient C see *An Introduction to Engineering Fluid Mechanics* by J. A. Fox, published by The Macmillan Press, London, second edition, 1977.

Answers to Problems

Chapter 2

2.1 (a) 28.32 mm Hg; (b) 8.5 mm Hg; (c) 19.82 mm Hg; (d) 22.0°C; (e) 23.7°C.

2.3 (a) 29.99 in.; (b) Cubley 29.76 in.; Biggin School 41.97 in.

2.4 39.1 mm.

2.5 (i) 29.8 in.; (ii) 31.2 in.; (iii) 27.0 in.

2.7 (b) About 1951; (c) assuming earlier period correct, increases it from 279 to 328 mm/year.

2.8 Some evidence of cyclicity, but longer record needed to show it conclusively.

2.9 (a) 0.136; (b) 7.34 m/s.

2.10 (a) Once in 4 years on average; (b) 23.8×10^3 m^3.

2.11 At X, 75 mm; at Y, 92 mm.

2.12 (a) 20 mm Oxford; (b) 64 mm Kumasi (if equation 2.4 derived for U.K. applies in Ghana).

2.13 481 mm; no.

Chapter 3

3.1 Amsterdam 4.0 mm/day; Seattle 0.05 mm/day; Khartoum 6.2 mm/day.

3.2 2.5 mm/day.

3.3 (a) April 5.24 cm; November 1.01 cm; (b) June 11.67 cm; October 3.70 cm.

3.5 6.7 mm/day.

3.9 Annual E_0 = 966 mm, pan coefficient = 0.85, net annual loss = 1.367×10^3 m^3/km^2/day, evaporation 136 mm less in July at 40°S.

3.10 14.076×10^6/m^3.

Chapter 4

4.3   4.1 mm.
4.5   (a) 53 mm; (b) 9 mm.
4.8   CWI = 129; compare with figure 4.9 value 125.

Chapter 5

5.1   Canal A 2.4 $m^3/m/day$; canal B 3.1 $m^3/m/day$.
5.2   kH = 2.32×10^{-2} m^2/s; R_0 = 1968 m; Q_0 = 0.044 m^3/s.

Chapter 6

6.1   4085 m^3/s; n = 0.032.
6.3   2560 m^3/s at 50.46 m.
6.4   (b) 0.705 $\times 10^6$, 1.85 $\times 10^6$, 0.1 $\times 10^6$ m^3/day; (c) 18.14 $\times 10^6$ m^3;
       (d) 5.14×10^6 m^3.
6.5   Additional storage 3.90×10^6 m^3; spillage (future) 12.01 $\times 10^6$ m^3; (now)
       21.37 $\times 10^6$ m^3.
6.6 46.8×10^6 m^3.
6.8   2124 m^3/s; n = 0.016.
6.9   (a) 74.2 l/s; (b) reservoir is drawn-down by 306 $\times 10^3$ m^3; (c) 765 $\times$
       10^3 m^3.
6.10 Capacity 65.73×10^6 m^3; yield 6.98 m^3/s.
6.11 (a) Assuming mean flow at 30% exceedance P = 9.8 MW; E = 66 GWh;
       (b) 227 m^3/s (using techniques from chapter 9).
6.12 (a) 3.5 m^3/s; (b) 75.0 $\times 10^6$ m^3; (c) 51 $\times 10^6$ m^3.
6.13 P = 1.2 MW (at 40% exceedance), E = 7390 MWh.

Chapter 7

7.1 Q_t = 126 $e^{-0.0231t}$; Q_{120} = 7.9 m^3/s.
7.2 Q_t = 120 $e^{-0.0392t}$; Q_{60} = 11.4 m^3/s.
7.3   Point N about 33 h.
7.5   76 m^3/s assuming same Φ index and a baseflow of 4 m^3/s.
7.7 Q_p = 705 ft^3/s; t_p = 4.1 h.
7.8 Q_p = 7.9 m^3/s; t_p = 4.0 h.
7.9   36.3 m^3/s at hour 7.
7.10 Point N is at 54 h; runoff volume = 557.3×10^6 ft^3; net rain 2.0 in./h:
       very severe for U.K. $T_r > 100$ y.
7.11 923 m^3/s at hour 6.

7.12 1332 m³/s at hour 6.
7.13 614 m³/s at hour 7.
7.14 687 m³/s at hour 9.
7.15 84.2 m³/s.
7.16 115.5 m³/s at hour 11.
7.17 (a) 125.4 m³/s at hour 12.

Chapter 8

8.1 At hourly intervals: 0, 2.3, 6.9, 22.2, 42.1, 70.8, 95.6, 115.6, 126.6, 128.9, 125.9, 116.8, 106.3, 93.8 etc.
8.2 137 m³/s at hour 32.
8.3 353 m³/s.
8.4 At hourly intervals: 0, 3.5, 9.8, 19.8, 37.4, 56.6, 72.7, 88.4, 98.0, 100.2, 94.0, 84.1 etc.
8.5 At 3 h intervals: 0, 5.5, 23, 46, 64, 75, 83, 87, 88, 87.5, 85, 82, 78, 72.5 etc.
8.6 250.4 m³/s.
8.7 Q_p = 428 m³/s at t_p = 8 h: net rain ≈ 1.4 cm/h.
8.8 At 3 h intervals: 6, 6.1, 3.9, 4.4, 16.4, 38.6, 63.2, 80.4, 91.1, 98.2, 88.7, 74.2, 60.3 etc.

Chapter 9

9.1 Q_{20} = 6360 m³/s; P = 18.6%.
9.2 (i) 37.0, 40.0 in.; (ii) 37%.
9.4 Q_{200} = 3220 m³/s.
9.5 (a) 68.04, 71.9 in.; (b) 0.401, 0.642, 0.871; (c) 0.395.
9.6 20%.
9.7 (a) 8.31 × 10⁶ m³; (b) 9.6%.
9.8 $\bar{Q}$ = 51 m³/s; Q_{200} = 163 m³/s: with safety factor (say 2.3) design Q = 375 m³/s.
9.9 Q_{100} = 140 m³/s; P ≈ 2%.
9.11 Q_{50} = 2799 m³/s.
9.12 (a) Q_{10} = 10.7, Q_{100} = 20.8, Q_{500} = 31.5 m³/s; (b) 3.8 m³/s.
9.13 71.9 m³/s at hour 9.
9.14 (a) From plot Q_{100} = 405 m³/s; (b) Gumbel Q_{100} = 392 m³/s, Q_{400} = 482 m³/s; (c) 22.6%.
9.15 5% in any year; 19% in any 4 years.
9.16 6 | 2, 40, 41 | 3, 10, 17, 25 | 1, 7, 11, 15, 20 | 18 | 19, 26 | 4, 27, 30 | 14 | 31, 33, 36, 37 | 4, 8 | 38 | 5, 33, 37 | 26, 28 | 22, 23, 24 | 13 | 7, 34 | 11 | 35.

Index